Swifts and Us

Swifts and Us

*The Life of the Bird
that Sleeps in the Sky*

Sarah Gibson

**WILLIAM
COLLINS**

William Collins
An imprint of HarperCollins*Publishers*
1 London Bridge Street
London SE1 9GF

WilliamCollinsBooks.com

HarperCollins*Publishers*
1st Floor, Watermarque Building, Ringsend Road
Dublin 4, Ireland

First published in Great Britain by William Collins in 2021

2022 2024 2023 2021
2 4 6 8 10 9 7 5 3 1

A catalogue record for this book is
available from the British Library

ISBN 978-0-00-835063-5

Typeset in Berling LT Std by
Palimpsest Book Production Ltd, Falkirk, Stirlingshire
Printed and bound in Great Britain by CPI Group (UK) Ltd, Croydon

MIX
Paper from
responsible sources
FSC™ C007454

This book is produced from independently certified FSC™ paper
to ensure responsible forest management.

For more information visit: www.harpercollins.co.uk/green

For David

CONTENTS

I

FRAGILE BONES

It is commonly believed that if you find a grounded swift, you should pick it up and throw it in the air, preferably from an upstairs window. This is one of those scraps of information that you absorb without thinking and which, like many other such unquestioned beliefs, is actually wrong. Yet this is what I did – and it did not help at all.

I had advertised an evening walk each week of the summer around a small town in Shropshire, to find where swifts were nesting. For thousands of years we have shared the buildings we live in with assorted wildlife. Nooks and crannies within the weathered masonry of our homes have sheltered many different living things; the walls between us and the outside world are porous. It did not happen intentionally but was a consequence of the building materials available. Today we can so easily make them airtight that we run the risk of losing our wild companions. Without thinking, we are shutting nature out, severing yet another link between ourselves and the wild world.

The contemporary zeal for sealing buildings against all elements and 'intruders' is thus depriving swifts of the crevices they may have nested in for centuries. A breeding swift will return to the same hole in the same house in the same street where its seldom-used feet touched down the previous year. This is what instinct impels it to do: to stick to the map in its memory; to navigate back to the exact place it has claimed, won and defended in order to rear its young. Unfortunately, growing numbers of swifts return to find their holes blocked off.

There are still dozens of swifts around town, but I know there are fewer than there used to be. People have told me of places where they used to see them, such as the Victorian Assembly Rooms, reincarnated as a nightclub for many years then converted to apartments with a renovation that left no holes for swifts. It is the same story in towns and cities all over the northern hemisphere. When they disappear from their old haunts, house by house, street by street, life ebbs away.

Roof renovation can easily be swift-friendly but it rarely is. In my home town I decided to find out which buildings were used by the birds, then to try and ensure their holes were retained when scaffolding appeared, before renovation work started. To my amazement, around a dozen people joined me on the first outing, fellow enthusiasts keen to find out more about this wondrous bird and its habits.

An hour before sunset is the best time to look: parent swifts will be snapping up the last insects of the day, a late supper for their nestlings. We scan the skies, listen for high-pitched cries and follow. Always on these evenings we are led into the older parts of town, along streets of Victorian terraces, above shops in the centre, down back alleys of old warehouses and beyond. Ventilation used to be regarded as essential in building construction, to prevent timbers from rotting. This ensured plenty of cavities for swifts and bats. With the advent of new materials, such as concrete and plastic, buildings started to be sealed off from the air, plastic soffits and fascia boards replacing wooden ones, with little chance of gaps opening up. Damp courses prevented moisture entering, and building regulations were drawn up, stipulating that ventilation holes must be bird- and insect-proof. Inevitably, this brought serious consequences for swifts and other hole-nesting creatures.

Discovering nest sites takes time and patience, but is endlessly interesting. Much of our time is spent simply watching the wild antics of the birds as they screech above the streets, dark crescents of energy. Every week until the midsummer solstice we set off a few minutes later, thereafter a few minutes earlier, hoping to catch sight of the parent birds as they return with the chicks' last meal of the day, flashing into crevices under eaves in the blink of an eye. Only when the bats start to emerge do we head home; this is a crossover moment after which you seldom see a swift.

These outings open my eyes to a side of town I have barely noticed before. We look up, scanning the eaves, searching for the tell-tale signs of thin whitewash that stains the walls for a few inches below their nests, watching for a swift to make its sudden return as it swoops back into its dark hole. It is like a treasure hunt; each new nest site raises a cheer. One or two in an evening is as many as we can hope to discover. Sometimes people tell us of their sightings: 'Look up at the third brick to the right of the drainpipe above the opticians on Willow Street,' writes one. 'They're nesting at the back of Betfred,' says another. Gradually, our knowledge of town grows a new layer, the elaborate mouldings and previously unregarded architectural details appreciated as we tune in to the swift's world, high above the shops and pedestrians walking the streets.

Staring up in the direction of top-floor windows inevitably attracts attention, especially when using binoculars. People stop and ask us what we are doing, which is good as it gives us a chance to talk about swifts. We meet others who share our passion. Nick, who works in an old mill in the crumblier end of town, emails me to say the sky is full of swifts around his building. We find it down a back lane. At the top of an iron fire escape a door is open. Forty steps up I call out and he appears, back from a run and eating his supper. It is a fantastic place to watch swifts, he says, the town's roofscape below and big skies all around. Several pairs are nesting under the eaves, but he thinks fewer birds have returned in recent years.

We call on Jas, a sculptor, who contacts me to ask if his home would be suitable for a swift nest box. He lives in a single-storey, brick-built, Victorian industrial building, imaginatively converted to flood his studio with light. His backyard has been transformed into a garden with a fountain, greenery and a bird feeder. On the adjacent wall of a taller building, he is planning to grow climbing plants. And put up a swift box or two. We talk about the possibility of him doing a swift sculpture; he already has several artworks in town: some naked acrobatic drinkers outside a pub, and a man leaning out of a window, arms stretched to catch an escaping chicken, above a fast-food shop.

We have seen swifts slipping under the roof of a redundant Victorian chapel. A few weeks later, scaffolding goes up and we worry. This splendid old building has been bought by an Italian restaurateur and restoration is under way. I speak to Paolo, the owner, and he expresses concern – he does not want to do anything that will harm *i rondoni*. Fortunately, the roof is found to be in good shape; no renovation work is necessary. Paolo is also interested in putting swift boxes up in the tower of his house, an imposing nineteenth-century Gothic building, perched on a hill overlooking the town.

Our last outing takes place towards the end of July, a warm summer evening during a spell of hot weather. Sun, blue skies and enough rain to keep things fresh have ensured an abundance of insects at ground, tree and aerial levels. It has been just right for the swifts; they could hunt from dawn till dusk to feed their growing chicks.

The sky seems quiet. A massive thunderstorm had broken over the town a few days earlier and it looks as though some of the swifts have already left. More than a third of the birds hurtling around in mid to late summer are immature birds, and their sojourn in Europe is even briefer than that of the breeders. We meet as usual in the square, by the statue of a stocky shepherd and his sheep. Where to go? While they may not have been nesting, those immature birds have given us clues as to where to look, screaming around the active colonies and searching for nest sites to claim and use in following years.

Our decision is made when one of our number, Andrew, takes a piece of paper from his pocket. He has received a message from an elderly friend, a man who worked on the nearby railway until its closure in the Beeching era. The note tells us to go to the Cambrian Railway Museum, formerly a goods shed, where swifts have nested for at least sixty years and where, he says, we will find a swift in a cardboard box. A cat had brought it in, alive and apparently unharmed.

The first flight of a fledgling swift is always perilous. They have one chance to get their wings together and go. The urge to fly is instinctive, but it still requires courage to take the plunge. Fully-fledged birds will shuffle to the entrance of their nesting place, peer out, teeter on the edge . . . then turn around and head back to the dark safety of their hole. The building had been used by two pairs of swifts this year and it sounds as though this casualty might have been from one of their broods. The shed is single storey, so fledgling birds have little more than a second to get flying or crash to the ground. Despite this disadvantage, generations of swifts have successfully negotiated its relative lack of height.

Our railway foundling is sitting in a shallow box with its incredibly long, narrow wings folded like crossed scim-itars. Andrew spots a mite crawling away from its eye. 'I've heard the best thing to do is throw them in the air from a good height,' says someone. This is what Ted Hughes did; he had written about it in his much-quoted swift poem. And no, it had not worked out well for his swift. Still, this is the accepted wisdom and we decide to give it a try. We pick up the bird, still in its box, and take it across the road to a house belonging to a friend of Andrew's. She lives in a flat on the second floor – ideal, we think. When we arrive, she and her young grand-daughter are piecing together a puzzle at the table, but they welcome us in and happily agree to let us get on with our reckless mission. She raises the sash window as high as it will go. Gently, Andrew lifts the bird from its box and launches it upwards over the street. Its wings open . . . and it belly-flops down to the tarmac below. If it was not injured in its first fall, it very probably is now.

On the ground, several of our group watch as the swift tumbles to the road and quickly scoop it up, fortunately still very much alive. Back in its box I carry it away; it turns its head like an owl and looks at me. Despite the wretched circumstances, this encounter is something to treasure. Never before have I seen a swift so close, so still. Wildlife has a way of vanishing before you get a chance to look at it properly, and with a swift, that most aerial of birds, you rarely get more than a passing glimpse. Well developed, with long wings, it has the markings of a juvenile: the finest white line along the outer edge of its primary feathers. Young swifts often have white foreheads too. This one is mostly brown but with a faint, pale edging to the feathers on its head, overlapping like fish scales for perfect aerodynamics. Around the corner of its eye is a black fringe of tiny, eyelash-like feathers. Nightjars, a not-so-distant relative of swifts, have these too. They protect their eyes from aerial debris and insects.

White fringes on the feathers of a juvenile swift.

For the swift, though, this is grim. Two crash landings in one day and a near-death experience when a monstrous furry animal fifty times its size had snatched it up in its teeth.

What to do now? A young lad and his girlfriend come along and look into the box, curious. I tell them its story. 'There's a twenty-four-hour vet down the road,' he says. 'They might help.'

The blind is drawn over the door but, seeing a light switched on, I call through the letter box. A young veterinary nurse opens up and lets us in. Instantly, the swift becomes a subject of administrative procedure, its details logged on the computer. 'Name?' she asks. 'Jonathan,' says Andrew, instinctively.

Naming it after an eighteenth-century literary Swift certainly bestows dignity on our bird but I have always been uneasy about giving wild creatures names; the practice seems to tether them to the human world. No sooner has this bird fallen to Earth than its identity is altering. What I want is to restore it to its aerial realm nameless, a swift among swifts.

The bird is taken away and we head home. Straightaway, I turn on my computer and look up Swift Conservation's advice on a grounded swift: DO NOT THROW IT IN THE AIR. Yes, that is a lesson we have learned, but what folly to have done this.

I lie awake, wondering how it is faring and whether it is badly injured. It had seemed lively, though, so I remain hopeful. Perhaps, I think, I can look after it. I had found detailed information on the internet about how to care for and feed young swifts, and a strong desire takes hold to see whether, given appropriate care, it might recover and fly. I have booked a couple of weeks off work without many plans, so I have time, I think.

The next morning I arrive at the veterinary surgery at eight-thirty. The vet on duty has checked the swift over and found no injuries, so I offer to take it away. The nurse looks doubtful but I show her my freshly printed pages on swift care. She sees my determination and fetches it from a back room, tucked into an empty latex glove box.

The swift looks terrible. Its eyes are closed and its body limp, but it sits, as before, perfectly symmetrically with its elegant wings crossed. I take it home and put it down gently in a recycling box on a towel, settling it into the spare room upstairs with two closed doors between it and my cat, Spindle. Its eyes do not open all day, nor does it move. I give it water, soaked onto a cotton wool bud, stroked along the side of its beak. If the water is dripped onto the top it would pour straight down its nostrils. The swift opens its beak a little and swallows so I repeat this every hour or so. Bright, spinach-green droppings start to appear.

I do give it a name of sorts: swiftling, a term of endearment that doesn't need a capital letter and has no inference of gender. The plumage of male and female swifts is identical; only when they are mating or egg-laying can you tell them apart, though it is known that male and female birds make up a duet of calls: that *sweree* sound is actually produced by a male and female bird calling alternately, and that there are subtle differences between male and female calls. But my swift is by itself and silent, so swiftling remains an 'it'.

The next morning it seems better: eyes open and a slight turn of the head. In my hand it feels stronger, not limp any more. I stroke its head and throat. Swifts like this; they preen each other when nesting and it relaxes them. Tremulous when first picked up, it soon grows calm. I give it more water and offer it a headless, legless cricket. These, I had read, are the best food to give a swift. My local pet shop had recently put up a sinister sign in the window: *Live Food Available*. Unsure quite what this might include, I had enquired about insects.

Crickets come packaged like grapes, in clear plastic cartons. Inside, they hunker down among the hummocks and hollows of egg-boxes, nourished by a sprinkling of bran. Most of them are destined to be snapped up by pet reptiles, but they also make excellent food for insectivorous birds such as swifts.

First, though, you have to remove the indigestible parts: heads and hairy legs. I have to grit my teeth for this bit; I am a vegetarian and prefer not to kill things, though I can be ruthless with slugs when they trail across my kitchen surfaces at night. But I have to think from the swift's point of view; their diet is composed of insects and spiders, aerial fodder blowing about the sky. So crickets it has to be and I soon become adept at the necessary snipping.

To start with, my swiftling will not accept these delicate offerings, its beak staying resolutely closed. Fearing its imminent starvation, I turn back to the internet for advice. Instantly, I find a German video with subtitles showing me what to do. *Use a blunt, rasped fingernail to carefully open the pecker.*

I do as instructed, first wrapping the bird in a soft cloth to keep it calm. Next, I open its beak, whereupon its tongue shoots out, a strong, pointed thing. When a parent swift feeds its chicks its whole head goes into the nestling's mouth and the food is deposited far back. I have to do something similar with a pair of eyebrow tweezers, carefully avoiding touching the soft skin inside its mouth. With my index finger I do my best to keep its beak open, which the swift seems determined to close. Finally, I get three crickets in, though it seems a small victory when I learn they need to eat at least sixty per day.

A parent bird delivers food to a nestling.

I am not happy with my tweezers, alarmingly sharp instruments that could easily inflict damage on the delicate tissues of its mouth, but I can't find blunt-ended ones in the shops. I ring Andrew, who is a retired eye surgeon. He has a set of surgical tweezers, including some round-ended ones. With these and growing confidence, I manage to feed the swift thirty crickets and it grows livelier, its droppings whiter. I fold a towel over the side of its box and soon it climbs up; swifts can rest vertically, like bats. Their hook-clawed toes are ideally adapted for this: two pairs set at an angle to each other, enabling them to cling onto walls and rock faces.

The swift grows used to me feeding it and I can feel

its hunger as it almost swallows the finger that holds its beak open. Between each mouthful of cricket I stroke its white throat. To preen it properly I use an artist's soft paintbrush; a bird's feathers need to be in perfect order for flight.

For two days I am in London and Andrew takes over. With his deft surgeon's hands he is quick to get it eating – this time with a pair of yellow plastic tweezers, gentler even than the round-ended steel ones. It tips its head right back, completely at ease as he strokes it.

Back home I weigh the swift as best I can on my kitchen scales with its old-fashioned brass weights. One ounce and . . . two 5p coins. In metric measurements I make this 34.85 grams. Weight is a good indicator of a fledgling's readiness to fly. Nestlings can reach around 52 grams, fattened by a dozen meals a day, but then lose weight through exercise. This involves doing press-ups with their wings to strengthen their muscles, pushing their whole bodies up, balancing on their wing tips. By the time they leave the nest their weight will have dropped to about 42 grams; if they are too heavy they will plummet to the ground. This swift, however, went without food on the day of its crash to the ground and has eaten little for a day or two afterwards, so it needs feeding up.

For the next week or so, my life revolves around the bird. Six or seven meals a day, ten crickets at a time. Once I knock the open container and a dozen of them leap out. I scramble to retrieve them. Spindle pounces more effectively, trapping them under his paw. One creeps behind the skirting board and he listens for hours for the silent

Andrew Tullo and the rescued swift.

scuff of cricket feet on floorboards, crunching it up when it finally emerges.

The nutritious crickets work their magic and the swift puts on weight – another 5p piece on the scales. Its plumage shines; in the garden, sunlight reveals the iridescence of its feathers, a purple gleam transforming its dark brown colouring. Healthy though it appears, it still shows no inclination to fly, no restlessness, no press-ups. On the palm of my hand I hold it up and notice one of its wings slip down. I do my best to put this out of my mind.

By now it is the beginning of August and all the local swifts have gone, probably halfway to Africa. Will I have to make myself a pair of wings and leap up and down the garden to show it what to do? I speak to a swift carer in the Midlands. 'Don't worry,' she says, 'there's still plenty of time – your swift will know when it's ready to go.' She is right, of course. A swift knows these things instinctively; I have to give it time.

Two and a half weeks after we found it, the bird is still in my spare room, shuffling across the carpet or hanging vertically on the fuzzy material of my car boot cover, a barrier I have put up to stop it disappearing into the dust and clutter under the bed. Once I come in to find it sitting comfortably in the middle of the bed, and sometimes it clings to my jumper, hooking its claws into the wool. My swiftling's instincts are clearly not telling it to fly.

Finally, I take it to a wildlife rescue centre in south Shropshire. No name is requested here; the veterinary nurse simply writes 'swift' on her list, which already includes rabbit, pigeon, sparrowhawk and house martin.

She looks at it carefully and expresses concern about its shoulder. A full examination reveals a broken bone; it is mending, but my bird is damaged, its exquisite flight equipment faulty. I will never know whether it was the initial fall from the nest or that well-intentioned throw from the window that caused the injury. The first we could not have prevented, the second was a mistake made in ignorance, never, ever to be repeated.

A bird that lives its entire life on the wing, that will fly several million miles in its lifetime, must be in perfect physical shape. This is no longer a possibility for my swiftling. Its brief life is over.

2

AWAKENING

One swallow may not make a summer, but the first screech of a swift irrefutably rings in the season, wildly announcing the promise of sunlit days and long, warm evenings. Not everyone looks up into the skies, not everyone is tuned into birdsong, but for those who are, the arrival of the first swifts at the end of April or beginning of May is a thrilling moment.

Still, it took me forty-odd years to notice them. Did I grow up with cotton wool in my ears? In the 1960s and 70s there were probably twice as many swifts as there are today, so they would have made one hell of a screech.

The wild end of the garden was our playground: under the tangle of bushes I made dens with my sister; in the more tenderly nurtured areas I pulled stamens out of golden azalea flowers to suck the sweet nectar, and picked sawfly caterpillars off roses, sixpence a dozen. One summer I spent days searching hedgerows for birds' nests and wondered at the intricate construction of a blackbird's woven cup and the mossy secrecy of the wren's spherical home.

Mostly, though, I sat in my room reading the books I had borrowed from the library: three a week, using my green junior member tickets. Later, we got a dog, a bouncing golden retriever who needed a lot of walks, so she got me away from my reading, out into fields and woods. Still, I did not notice birdsong. House martins, yes. They nested right outside my bedroom window, chattering excitedly in their mud bowl nests, so I couldn't miss them. Yet the Sussex Weald, rich in woods, old pastures and hedgerows, would have been alive with song: multitudes of warblers doing their stuff by day, nightingales filling the air with melody by night. I must have heard all this but remember none of it.

Seared into my memory, though, is the destruction of hedgerows in the fields around our home in the 1970s. Agricultural intensification was in full swing and farmers were being given grants to grub out hedges in the name of efficiency. This was my brutal awakening to the reality that nature is constantly under threat, that the landscapes we love are ever vulnerable.

Swifts came to my attention briefly when I was twenty – in a book. Searching for a birthday present for my father, I stumbled on Derek Bromhall's *Devil Birds*, the vivid story of Oxford's swifts, which accompanied his film, aired by Anglia Television in 1980. I read the book before giving it away, and discovered for the first time the extraordinary miracle of a bird that spends virtually its entire life in the air.

I married a birder, so my ornithological horizons broadened; his study was lined with bird books and he kept detailed records in black notebooks of all that he saw. At a motorway service station in Cumbria I listened to a willow warbler for the first time, and that sweet cascade of notes has been a joy every spring since. But we lived parallel lives and birds were seldom shared territory. The marriage lasted twenty years but never thrived. When I left, I mentioned the fact that even the house martins that had nested at the front of the house for seventeen years had left us, shunning an unhappy house. He, too, had taken this as an ill omen.

From a rural house in Shropshire, on the border with Wales, I moved into Oswestry, a small market town a few miles away. Straightaway, I knew that I wanted to share my house with swifts. A long-dormant seed of awareness had suddenly grown into a passionate desire to be close to these birds; I was a town creature now, but that did not have to mean leaving behind the wild things I loved about the country.

Swifts are living symbols of wildness. They are the most aerial of birds, staying on the wing for years at a time, rarely touching ground for a single moment. They catch all their food in the air: aphids, flies, spiders, beetles, mayflies, even small moths and dragonflies, whirled into the sky, carried on the wind. Swifts drink and bathe on the wing, sleeping and even mating in the air. They fly closer to the sun than any other bird, feeding and sleeping at altitude. This is their realm, the world to which they are supremely adapted; Icarus, be green with envy.

Yet there is one thing the sky denies them: a place to rear their young. For this, they must descend like mortals and behave as other birds do: seek out a safe hideaway, build a nest and breed. The places they choose to nest in could not be more different from the boundless, aerial world. Not for them the wind-blown treetops that a rook, heron or mistle thrush would choose, but a dark, tiny cavity, sheltered from rain, storm and predators. And a place their feet can shuffle along, where their eggs will not roll around. For most European swifts, this means a hole under the eaves of a house or crevices in the walls of churches, old factories, warehouses and castles in cities, towns and villages. Few swifts now nest in natural habitats in Europe; they have adapted overwhelmingly and very successfully to human architecture.

I knew they would be in Oswestry and, as if by instinct, I was tuned in, alert to the earliest returning birds. Their calls had registered in my memory without my even noticing, the soundtrack of summer, heard in towns and villages all over. Often now, I hear their calls in films – *sweree* – a uniquely evocative sound. That May I first heard them through my kitchen window. I shot out to watch a pack of screamers speeding overhead, black bolts of wildness.

With all my heart I welcomed these birds back, but I wanted to do something practical, too. My new garden, a dull rectangle of lawn and decking, had great growing potential. I would have a pond, a hawthorn hedge, angelica, sweet cicely, thyme, lavender, a patch of wildflower

meadow with yellow rattle and knapweed, ivy and honeysuckle up the walls; every last scrap of space could be
brought alive. An abundance of insects would, by some
earthly miracle, move in – spiders, flies, froghoppers and
bees – many of which would be snapped up by bigger
insects and birds, while some would float up into the
sky and feed swifts and martins. Meanwhile, I would
ensure that the holes at the ends of the gables would
stay open; the house would have places where swifts
could nest – sparrows and starlings too, if they would. I
determined to do everything I could to lure them under
the eaves.

With help from the new man in my life, the transfor-
mation began. He would arrive with a crowbar to lever
up the decking and a chainsaw to fell the leylandii, swing
a ladder against the house to put up swift boxes, bring

evening primrose and borage plants, and buckets full of broad beans and blackcurrants from his allotment for supper. My son astounded me by digging a deep hole for the pond, with an enthusiasm for spadework unseen since childhood days on the beach.

Already, swifts were tantalisingly close, racing over the rooftops and calling to each other, high in the sky over my garden. Then I saw that they were nesting in my neighbour's house, slicing down the narrow gap between the buildings, turning sharply under the eaves. If I played the swift calls loud enough, their offspring might return the following year and notice the new boxes on the other side of my house.

Making my own patch bird- and wildlife-friendly was the first step, but I wanted to know how swifts were faring where I lived and across our fast-changing world. I feared the worst. I had witnessed a slow loss of swallows around my old home along with the disappearance of house martins. So I was braced for the news that swift numbers were falling steeply. Over the last twenty years the UK breeding population has halved.

Yet grim though this statistic was, what I uncovered was a story of hope, of people taking simple, practical steps which would make a difference; determined people doing their level best to ensure that swifts remain part of our summer, screaming through our skies, making their scratty little nests alongside us for hundreds of years to come.

My curiosity about these strange birds was growing. I

wanted to learn about their lives, find out how people had gradually unravelled their story over centuries, and how they had evolved from a tree- and cliff-nesting bird to one almost entirely associated with buildings and the urban environment.

3

ANCIENT SURVIVOR

It sometimes happens on a mountain in thick mist: a gust of wind will suddenly blow the clouds apart, giving you the briefest glimpse into the valley below. It is like this with fossils: they are snapshots into ancient chasms of time.

These fragments of petrified creatures give us tantalising clues about the history of life on Earth. The house in the hills where I previously lived was built of local limestone, its walls studded with the fossils of ancient sea creatures, while the step down into the pantry from the kitchen, a polished slab, held the white, tubular shapes of crinoids, vivid against the dark stone. Here, in landlocked Shropshire, more than seventy miles from the sea, was dizzying evidence of another time and place. This limestone had formed in a shallow tropical sea more than 300 million years ago, shifted by the continental drift that keeps the earth's landmasses in constant, imperceptible motion.

Crinoids were once known in folk memory as 'fairy money'. Though philosophers have speculated about such fossils and their connection to the marine environment for more than 2,000 years, it is only relatively recently that their significance has begun to be clearly understood. The word 'palaeontology' first appeared in 1822 in France, coined to describe the study of ancient living organisms through fossils. A surge in geological exploration and study over the next century unravelled many of the mysteries of prehistoric life.

The discovery of a fossilised feather in Germany in 1860, followed by a near-complete skeleton three years later, advanced our understanding of the origin of birds by a leap. The skeletal creature was named *Archaeopteryx*, meaning ancient feather or wing in Greek. It had been alive some 150 million years ago, in the late Jurassic period. With its sharp teeth and long, bony tail, it vividly resembled a dinosaur but its feathers raised questions. In 1868 Thomas Henry Huxley was the first scientist to suggest it could be an evolutionary link between dinosaurs and birds – a transitional creature.

Further similarities between birds and dinosaurs emerged as more fossilised remains were unearthed around the world. Dinosaurs had hollow bones as contemporary birds do; they made nests and laid eggs resembling those of birds, brooding them in a similar way. Endearingly, scientists also cite the habit of both dinosaurs and birds of sleeping with their head under their forelimbs or wings – perhaps to keep their head warm – as evidence of their shared ancestry. Yet it would be another 100 million years before the first swift-like birds evolved.

Fossils have emerged from quarries and land slippages across the northern hemisphere, revealing evidence of our swifts' forebears; they have flown through our skies for nearly 50 million years, gliding and beating the air with their long, slender wings.

Some of the earliest evidence of primeval swifts emerged in the London clay in southern England, and has yielded a wealth of detail about the nature of this land in the early Eocene era, the geological period named from a Greek word meaning dawn of the new, and the time when many new types of fauna emerged. At this time, the land lay some 10 degrees further south and was vegetated with mangrove ferns, magnolia trees and other tropical vegetation, similar to the forests of Malaysia and Indonesia today. Fossils of tiny horses, 14 inches high, have been found, along with turtles and crocodilian reptiles. In the Fish-Tooth Beds of Bognor Regis and Warden Point on the Isle of Sheppey, along with other bird relics, two small humeri, the stocky wing bones of a primitive swift, were discovered in the 1930s. These were described in detail some forty years later by palaeontologists at the Natural History Museum in London. More remained of the Bognor bone, which became the holotype for a new genus, *Primapus lackii*, the first swift, named in honour of David Lack, whose pioneering study of swifts had been published in 1956. Scientists assigned it to the fossil swift family Aegialornithidae, similar to modern treeswifts, which are closely related to true swifts.

A few years later another bone emerged in Essex, in the rich fossil beds at Walton-on-the-Naze. Along with ancient fish bones, shark teeth and carbonised tree trunks, the fossilised remains of more than 150 bird species have now been found, some with teeth, revealing their dinosaurian ancestry. Among these petrified marvels, palaeontologists found a distinctive humerus (upper wing bone), from a species which would be named *Eocypselus*. It appears to be a primitive treeswift but also has some anatomical similarities with owlet-nightjars.

Evidence of early swift-type birds has emerged around the world too, fragments or even complete skeletons preserved over many millions of years. In the west of the United States, in Wyoming, Utah and Colorado, a series of dried-up lakes along the Green River has yielded a rare treasury of fossils, preserved in the sedimentary lime muds. Absence of oxygen within these fine-grained sands halted the process of decay, revealing the burgeoning life of the early Eocene period. There are even fossils of insects, their delicate wing membranes clearly visible, and leaves and ferns with the bite marks of nibbling insects, intact in wondrous detail. Thousands of fish, reptile and bird fossils have been unearthed, along with the first recorded bat, complete with its last undigested meal, eaten around 50 million years ago.

In one of the lakes in Lincoln County, Wyoming, a tiny fossilised bird skeleton was discovered with features recognisable in modern swifts: the stocky humerus and an elongated carpometacarpus (tip of the wing skeleton), both characteristic of the swift and its closest relative, the hummingbird. Described by palaeontologists in 2013, their investigations point to it being an early example of the Apodiformes order, which includes hummingbirds, treeswifts and swifts.

Named *Eocypselus rowei*, this 52-million-year-old bird is the forerunner of both swifts and hummingbirds. That sturdy upper wing bone would ultimately enable tiny, nectar-seeking birds to hover above flowers, flapping their wings fifty times per second, and swifts to reach greater speeds than any other bird in level flight: cheetahs of the sky. The shared ancestry of these birds, which few of us would ever have imagined, seems less surprising when you think of the exceptional aerial abilities of both.

By a miracle of natural preservation processes, some of its feathers were still intact. This bird had a crest, just as some treeswifts and hummingbirds do today, and a wing shape between that of a swift and hummingbird; the wings are shorter than those of contemporary swifts but longer than those required for the hovering flight of a hummingbird. Scientists – quite astonishingly – were able to identify melanin within the cell structures, indicating that the bird was probably black and possibly had an iridescent sheen, just like our swifts.

Perhaps the richest-known fossil bed of the Eocene period lies in a former oil shale mine in Hesse, Germany. The Messel Pit's value has been recognised by fossil-seekers since the late nineteenth century when mining began. However, when industrial activity ceased in 1970, the quarry narrowly escaped being turned into a landfill site. Following an outcry by local scientists and an international campaign to save it, the pit was purchased by the state of Hesse, designated a cultural monument and, in 1995, a UNESCO World Heritage Site.

Like the Green River formation, fossils at the Messel Pit preserved details of both plant and animal life. Scientists believe that volcanic gas released into the lake caused the sudden death of creatures, poisoning its waters and surrounding vegetation. Birds and bats flying over its surface may have been overpowered by the noxious vapours and fallen into the lake, drifting down into its murky, oxygen-poor depths, conditions hostile to putrefying bacteria. Entombed in mud that over millions of years turned to stone, their bodies were preserved.

That disastrous incident, around 47 million years ago, bequeathed us an extraordinary gift. The sheer diversity of species living in this once sub-tropical landscape is mind-boggling: among thousands of fossilised skeletons, evidence has emerged of an alligator, pygmy horses, gigantic mice, a flying ant with a six-inch wingspan, pangolin, aardvark relatives, countless numbers of insects, including jewel beetles still gleaming with colour, a long-tailed hedgehog-type creature, an early potoo (like a nightjar), a mousebird with stubby toes (found today only in sub-Saharan Africa, a small greyish-brown bird that scurries about the tree canopy in search of berries), nine pairs of mating turtles, doomed in the act of procreation, and more than 700 bats. As if by magic, some fossils were preserved complete with hair, skin and feathers.

More than sixty plant families have been identified, including examples of blossom, leaves, seeds and fruits. The humid climate gave rise to tropical vegetation, including ferns; cashew trees; an evergreen, nutmeg-like species; tree of heaven; incense trees and *Cyclanthus*, related to the Panama hat tree. Giant birds thrived during this time when there were no large predatory mammals – they became the hunters.

Of several hundred bird skeletons discovered, about fifty species have been identified. Among them, from the Middle Eocene, was *Aegialornis szarskii*, its long, crescent-shaped wings beautifully preserved and undeniably swift-like. This fossil, on display in the Senckenberg Museum in Frankfurt, has earned the distinction of being the earliest true swift discovered so far.

A fossil swift from 49 million years ago
in the Senckenberg Museum, Frankfurt.

4

SWIFTS AROUND THE WORLD

I have borrowed a book, heavy as a hod of bricks. This mighty tome is *The Illustrated Checklist of the Birds of the World, Volume I*. On its gleaming front cover is a finely drawn tree of life, branching out like a willow caught in the wind. More than 100 different birds are perched on or flying around the tree, grouped according to their order within the avian world.

To one side at the bottom are the Palaeognathae, a clade or group of birds. These include the most ancient species: flightless ostrich, kiwis and cassowaries, and forty-seven types of the secretive tinamou, ground-dwelling birds that live in Central and South America and which are capable of flight but prefer to run.

The second basal clade is the Neognathae, which includes the other 10,000 or so living birds. The group-ings are based on structural differences, including their jaws – Palaeognathae means old jaw, and Neognathae new jaw. On an upswept branch from low in the tree, several quite similar birds are perched. They resemble broken-off chunks of bark with big, dark eyes: nightjars, frogmouths and owlet-nightjars. Just beyond them, flying from the tip, is a swift.

A species is generally defined by reproductive isolation, each one mating and rearing young only with its own kind. This approach is somewhat complicated by the fact that individual creatures do occasionally interbreed with other species and sometimes produce hybrids, but these will be sterile. Indeed, it is estimated that 9 per cent of all bird species in the world have interbred outside their own species. However, such incidental couplings do not generally endanger the species itself, which retains its own distinct identity. While genetics is helping to clarify the orders and families into which birds are divided, it has not yet superseded traditional methods of taxonomy, based on plumage, skeletal structure, vocal sounds and so on.

A degree of complexity is thrown into the science of taxonomy by the very people who conduct it: human perception varies and one taxonomist may interpret information in a different way from another. Meanwhile, life forms are constantly evolving, so, for many reasons, the names, places and hierarchy of birds within avian genera are subject to periodic tweaking.

I idle through pages of the world's nightjars, revelling in their cryptic feathering – tawny, grey, brown and the odd fleck of white – rendering them all but invisible as they roost along branches or among dead leaves and fallen wood on the ground. Like their relatives, the swifts, they have bristles around their eyes, short legs and long, pointed wings. Their feet are small and little used for walking, but these birds too are agile aerial hunters, snapping up moths, beetles and flies into a cavernous gape. Feared by country folk for centuries, on account of their suspiciously nocturnal ways, their reputation for wickedness is immortalised in the genus name, *Caprimulgus*, meaning drinker of goat's milk. An endemic nightjar species in Sulawesi, with a buff-yellow collar, has an even more sinister name: *Eurostopodus diabolicus*, the Satanic nightjar.

Next come the treeswifts, the Hemiprocnidae. These elegant-looking birds have a hind toe that enables them to perch on branches, long tail feathers and, in some cases, a crest. The grey-rumped treeswift has a dark, fir-green sheen to its upper parts and brick-orange ear coverts. Whiskered and moustached treeswifts have bold white face markings. Like a bracket fungus, their nest is fixed to the side of a slender branch, a tiny half-saucer made of saliva, feathers and splinters of tree bark. A single egg is laid and incubated, the parent bird perches on the branch – the nest is too fragile to bear its weight – just its brood patch in touch with the egg.

Then I find what I am really looking for, the Apodidae family, eight glorious plates of bright, accurate illustrations of the hundred-odd species of swift living around the world. Bright, that is, in a swift-like way, though somewhat dull when compared with the jewel-like hummingbirds that follow them. Nearly all have brownish-black plumage; some have areas of white feathering, arranged in a spot or a band across the belly, nape or rump. A rare splash of colour draws my eye to the chestnut-collared swift, found in South America on the western side of the Andes, and in Central America from Mexico down to Guatemala.

The evolutionary hierarchy of the swift is quite complex, but soak up just a little and very soon you find yourself drawn in by the differences between genera and the astonishing lives and adaptations of these birds. First, take a deep breath and grasp the basic structure. Within Apodidae there are two sub-families. The first, Cypseloidinae, includes the primitive American swifts of two genera: *Cypseloides* and *Streptoprocne*. The second, Apodinae, has three tribes: the swiftlets (Collocaliini), needletails (Chaeturini) and typical swifts (Apodini). Each tribe then subdivides into genera, with distinctive but not always obvious characteristics. Worldwide, around 100 species of swift have been identified. They range over rainforests, mountains, rivers, towns and cities across most of the globe, absent only from the polar regions, southern Chile and New Zealand.

The names themselves are wonderful. Ancient Greek and Latin were never on my educational spectrum so I find myself sidetracked, exploring the meanings of the scientific nomenclature. The family name Apodidae describes a shared characteristic of all the swifts. Derived from a word meaning footless, it is not a literal description but refers to the fact that their feet are rarely visible and rarely used. Swifts do not perch, though they can cling to vertical surfaces with their powerful toes.

The nine species of *Cypseloides* (from a Latin word for

'swift') chiefly occur in Central and South America and have a strong affinity for water. They live in mountainous regions, nesting in deep river gorges on ledges darkened by overhanging moss or in crevices behind waterfalls; such places provide the eggs and nestlings with a high degree of safety from predators. Mud rather than saliva is used to bind nesting materials together. When the rains come, as frequently happens in the mountains, parent birds will roam great distances to find food for their young, feeding them just once or twice a day at dusk. The fledgling's first flight through a wall of water is a hazardous baptism into its aerial life.

The black swift is the most numerous of the genus, with a distribution that extends from the mountains of Mexico in Central America and the West Indies, up the west coast of North America as far as Alaska. Despite its long range, this is an elusive species, often feeding at high altitude and nesting in remote gorges and on sea cliffs. Intrepid ornithologists, fascinated by the mysteries of this bird, have scoured mountains, waterfalls and river gorges, and scaled slippery rock faces in their search for breeding colonies. Charles W. Michael, assistant postmaster at Yosemite National Park and a keen gorge scrambler and birdman, wrote of watching a black swift in 1926: 'Such bewildering speed, such co-ordination of mind and muscle. I stood there fascinated, trying to follow the flight of the wild winged thing. Somehow the thought came to me of a great winged spider suddenly gone mad.'

Great dusky swifts feed over the rainforests of South America and are abundant in Brazil. On the border with

Argentina they may be seen flying through the spray of the mighty Iguazu Falls, clinging to the vertical cliffs and slicing through cascades of water to their nests. Unusually for this darkness-loving genus, the great dusky swift will also sometimes make its mossy nest on ledges exposed to sunlight.

Great dusky swift flock in front of Iguazu Falls.

The five species in the *Streptoprocne* genus all have a white or orange collar contrasting with their generally brown plumage. The Greek word *streptos*, meaning collar or necklace, refers to this feature. The second part, *procne*, comes from the name of the woman married to Tereus, King of Thrace, who was transformed into a swallow by the gods when she and her sister fled after a violent and gruesome family episode. I am amused to see the ancient confusion between swifts and swallows perpetuated in taxonomy, the very purpose of which is to define the distinctions between species.

Among the *Streptos* I find my friend, the eye-catching chestnut-collared swift, though I am thrown by its inclusion here, as another of my books has it down as a *Cypselus*. It has been officially reclassified, along with its close relative the tepui swift, which glows even brighter: vivid orange around its throat and nape, all the way to its ear coverts. The females of both these species, however, are uniformly brown. Skilled swift-watchers recognise them in flight by their 'jizz'. In the same way that we recognise individual people at a distance by their distinctive walk or a tilt of the head, those who spend time watching birds will absorb characteristic features such as length of wing, flight style, depth of wingbeats, giving them the ability to identify species instinctively.

The biscutate swift, a typical *Streptoprocne* in many ways, with its broken collar of white feathering, along with a shield-shaped patch on its throat, defies the family preference for nesting near water by raising its young in dry caverns, where it weaves a nest of moss, lichen and dry leaves. How does it collect these materials? I imagine it flying near waterfalls, clinging briefly to vertical rockfaces to gather shreds of nesting material. These gregarious birds roost as well as nest in the caves, sometimes in huge numbers. In 1978 an estimated 90–100,000 were counted clinging vertically to the walls of a cavern (locally known as a 'swallow grotto') in the Serido region of north-east Brazil.

Next comes the Apodinae sub-family, with three tribes

and seventeen genera. The Chaeturini tribe includes the seven spinetails, found in Africa, Indonesia and the Indian subcontinent. Several are endemic to certain areas, such as the Madagascan and São Tomé spinetails. Their wings are broad, often curving to a point, hooked like an old-fashioned butter-knife. With a protruding head and chunky body, they have been described as 'flying cigars', ranging in size from 4–6 inches. Their tail feathers are stubby, squared-off or occasionally forked. Around ten spines protrude beyond the length of the vane feathers that make up the rest of the tail. Some nest in tree holes, and one, the lesser Antillean swift, was once recorded with a clutch of three eggs in an abandoned oven. Meanwhile, Böhm's spinetail, a sub-Saharan African species also known as the bat-like spinetail, makes its solitary, twiggy nest underground, on the walls of pits, shafts or wells. Like the majority of birds, it has anisodactyl feet: three toes pointing forwards and one backwards. It can swivel its backward toe, enabling it to cling onto vertical surfaces.

The *Chaetura* needletailed swifts live in the Americas. Carl Linnaeus, the father of modern taxonomy, believed these New World chimney swifts were swallows and classified them accordingly as *Hirundo pelagica*, in which category they stayed until 1825, when the ornithological authorities formally placed them in the *Chaetura* genus. This species breeds largely in the east of North America, while its close relative, Vaux's swift, is generally found in the west. Spectacular gatherings occur at dusk on migration to their wintering grounds in Central and South America, as the birds flock to roost, just as starlings do in Europe. Up to 35,000 Vaux's swifts have been counted, spiralling down into the old smoke-stack of a school in Portland, Oregon, sometimes watched by hundreds of people.

The loss of these old chimneys through redevelopment, along with urban annoyance about their mess leading to chimneys being capped, is causing big problems for the birds. They need these stopover points to break their long journeys. Chimneys are also used for nesting, almost exclusively so by the chimney swift. Vaux's swifts tend to nest inside old woodpecker holes – except in Colombia, where they too prefer chimneys. Old trees too are growing scarce with the destruction of mature woodland. Inevitably, both species are declining in numbers.

The *Hirundapus* genus of needletails has an oriental distribution. All four species have a distinctive white horse-shoe marking from the belly to under-tail, with dark, glossy plumage above. The tail shape varies according to whether it is fanned open or closed, making it round, pointed or something in between. This tribe includes the largest swift in the world, the purple-needletail, which lives in the Philippines and in north-east Sulawesi. Its lustrous purple body is just under 10 inches long and it has a wingspan of up to 24 inches. There is a surprisingly large variation in size between races in different areas of its range. In some parts of the Philippines it is persecuted for its habit of catching bees around apiaries. Curiously, the habits of this bird are little known, including its place of nesting and its call.

I look closely at the illustration of the white-throated

needletail, reputedly the fastest of the swifts, said to fly at an air-smouldering 105 miles per hour, though some have questioned the veracity of this figure. It looks surprisingly bulky for such a speedy bird, but its broad wings power it along with rapid beats. The tail spines are quite short; I compare them with those of the brown-backed needletail, which are significantly longer and longest of all at the centre, just like a packet of sewing needles. A disoriented white-throated needletail turned up in the Outer Hebrides in 2013. Their usual wandering grounds stretch from the wooded hills of Siberia to the foothills of the Himalayas in the breeding months, overwintering in eastern Australia and the Indian sub-continent, so this unfortunate individual was drastically off-course – much to the delight of hundreds of twitchers, who rushed to the Isle of Harris, only to see their bird crash into a wind turbine on its third day on the island.

Swiftlets, small swifts found in tropical and sub-tropical areas of southern Asia, the south Pacific islands and north-eastern Australia, are generally mouse-brown, with subtle variations, making them excruciatingly difficult to identify. They belong to the Collocaliini tribe, which is divided into four genera, with around thirty species. One has been given the dreary name 'uniform swiftlet' and is described unhelpfully as nondescript. A few are distinctive, such as the glossy swiftlet, with its gleaming dark blue plumage, and the white-rumped swiftlet; others can be identified only in the hand. Here, too, is the smallest swift in existence, the pygmy swiftlet, just 3.5 inches long, a Philippines endemic.

The name Collocaliini comes from the Ancient Greek words *kolla*, meaning glue, and *kalla*, nest. While many swifts use saliva as a binding ingredient in their nests, this is a particular feature of the swiftlets, especially the *Aerodramus* genus. Male and female birds emit strands of a gelatinous substance from salivary glands below the tongue, gradually building up a shallow dish of a nest, which hardens like fibreglass. Unfortunately named in recognition of humans' culinary exploitation of this species, the edible-nest swiftlet uses only saliva, resulting in a white nest, while the black-nest swiftlet adds some of its own feathers. These glutinous structures, built colonially in caves, have been harvested for more than 1,000 years for their supposed medicinal and nutritional properties. Marvellous powers have been attributed to the substance; it is said to promote long life, passion, energy, concentration and good skin – an elixir for all ills. Analysis has revealed that the saliva is rich in protein and amino acids, but there is a singular lack of evidence to support the extravagant claims of its special powers.

Mao Zedong regarded bird's nest soup as a decadent luxury, but trade in this traditional Chinese delicacy has boomed in recent decades. Sales of swift nests account for 5 per cent of gross domestic product in Indonesia – a stupendous $4 billion per year – and rising. To supply the apparently insatiable demand, purpose-built concrete nesting houses have sprung up in northern Sumatra, Vietnam and Thailand, with an estimated 40,000 in Malaysia alone. Most of the harvested nests are consumed in Hong Kong and mainland China.

Humans are strange creatures. Supposedly civilised, we think nothing of stealing the nest, the crucible of life that the swiftlet has spent at least thirty-nine days making in order to raise its chicks. It might seem odd, too, that we should choose to eat something made of spittle. Yet its perception as a delicacy and exorbitant financial value contribute to the nests' desirability just as much as the reputed health-giving properties and mythical associations. Bird's nest soup is the edible equivalent of the emperor's new clothes: tasteless and negligibly nutritious, but saturated with cultural value, it is transformed into liquid gold.

The swiftlets' natural nesting places are limestone caves, of which there are many along the rocky coasts of southeast Asia. They choose the darkest caverns, stretching deep into the cliffs, building their nests in vast colonies high on the walls and ceilings. The birds orient themselves in the darkness of the caves' tunnels and chambers using echolocation, generating a series of clicking sounds. It sounds 'like a thousand muffled typewriters', writes Edward Posnett in *Harvest*, in a fascinating chapter on the swiftlet industry.

It was from these caves that people originally gathered their nests – and still do – a dangerous undertaking but irresistible because of the money to be made. Hundreds of feet up, people erect ironwood poles up the cave walls, climbing in total darkness to gather the nests. In the 1980s and 90s, the price of edible nests in Hong Kong rocketed to US$1,800 per kilogram for black nests and $6,600 for white nests, causing a gold rush of collectors to descend on the caves. A harvest once gathered by local people in a sustainable way has now turned into a frenzied grab. In the Great Cave at Niah in Borneo, there were thought to be more than a million pairs of black-nest swiftlets in 1959. Since then, their numbers have crashed by 80 to 90 per cent.

Even the birds nesting on apparently remote islands are not immune. In the Andaman and Nicobar Islands of India, an 80 per cent population decline over ten years has been recorded, leading to the edible nest swiftlet being listed as critically threatened.

Purpose-built swift houses have, for many years, massively increased the trade in edible nests. This industry is hugely profitable but is showing fragility. Rapid urbanisation, forest fires and a rising population have taken their toll on Java, once the hub of edible nest production, and swiftlets, the golden geese of the island, are leaving. The tale is repeated throughout Indonesia and Malaysia: 'I saw rows of deserted bird houses, squalid flats surrounded by sewage and rusting cars, their tweeters blasting birdsong into the empty air,' writes Posnett. Borneo is the current hotspot for swiftlet farming, but for how much longer can it be sustained? Change is underway here too, with vast areas of rainforest felled and replaced with palm oil plantations, a devastating loss for many wildlife species.

The Australian swiftlet, the only one to breed in that country, has a unique incubation strategy. Like other swiftlets, it builds its nests in colonies in dark caves. Like others, it uses a combination of twigs, feathers and grasses glued together with saliva to create a cup-shaped nest, and, like others in the swift family, egg-laying appears to be stimulated by rain. However, the timing of its laying is utterly distinctive. Two clutches of a single egg are laid, the second one appearing three weeks before the first chick fledges. In this way, the nestling chick assists in incubation, enabling the parent to keep hunting for food. The Australian swiftlet is the only bird in the world known to behave in this way.

Then I turn the page and see a species I recognise: Alpine swift. Now, though, I am looking at it in a global context, for this jumbo jet of a bird has a vast range that takes in mountain ranges far beyond the Alps, from the Carpathians in Romania, the Taurus Mountains in Turkey, over the Atlas peaks as far as the Himalayas. And this is just its northern breeding range. *Tachymarptis melba* also breeds in South Africa and is resident year-round in the Indian subcontinent and Sri Lanka. Currently a scarce visitor to the UK, it has been suggested that Alpine swift is one of a number of continental bird species that might perhaps breed here in the future, as a result of climate change.

Swiss scientists carried out research into the behaviour of Alpine swifts using minute data-gathering tags with light and acceleration sensors to track the birds' latitude and longitude each day and to record activity levels. They were astonished to discover that the three individuals from whom they recovered the devices had stayed continuously on the wing for the six-month wintering period in southern Africa. Until then, evidence for such uninterrupted periods of flight had only been gathered for the common swift.

Formerly assigned to the *Apus* genus, the Alpine swift, along with the mottled swift, is now recognised as a separate genus. This classification is partly based on subtle differences in the foot shape of their nestlings. They also host different feather lice from *Apus* species. Recent genetic analysis supports their distinctive nature.

The Alpine swift's striking white belly patch, along with its large size and powerful body, makes it instantly recognisable. Its call is a rapid, staccato trill, quite unlike the scream of a common swift. Strongly colonial in its breeding habits, it nests on ledges or in holes on cliffs or tall buildings. These colonies may exist for hundreds of years. Mottled swifts, their close relatives and endemic to sub-Saharan Africa, invariably nest in crevices in granite cliffs – an apparently instinctive geological preference.

Finally, here is the *Apus* genus, encompassing twenty species. All of them have some degree of gloss in their plumage, and most, but not all, have long, deeply forked tails. Their toes are pamprodactylous but – here it gets complicated – functionally heterodactylous. What this means is that they have two opposing pairs of toes and the second and fourth have claws forwards, the other two backwards. The scientific nomenclature is appealing, reminding me of pterodactyls (a name that means 'winged finger') and the avian connection with dinosaurs.

Apart from Pacific, house and dark-rumped swifts, the *Apus* genus mostly lives in Africa – at least part of the time. Some, such as the Nyanza swift, are endemic. Nyanzas are a sub-Saharan species, similar but slightly smaller than the common swift. A huge colony breeds on the cliffs of Hell's Gate Gorge in Kenya, feeding by day over Lake Naivasha. Little swift, *Apus affinis*, is a highly successful member of the genus, with a wide-ranging distribution from sub-Saharan Africa to south-east Asia. Like the common swift, it has adapted to man-made habitats, both for nesting and feeding, and is abundant in towns and cities. The rainforest, though, is its optimum habitat and here it breeds all year round, two or three birds sometimes laying eggs in the same nest. One pair in Zanzibar laid six clutches over twenty-one months. In urban areas they sometimes come into conflict with house sparrows – just like our own swifts.

Another bird in this genus which, along with the little swift, has recently gained a toehold in Europe, is the white-rumped swift. The majority of its population lives throughout the year in Africa, but following its appearance in the Atlas Mountains of North Africa, it has expanded its range into Spain and Portugal, where it is a scarce breeder.

Formerly known as the mouse-coloured swift, the pallid swift has a scattered distribution, mostly in coastal areas. It breeds in both natural and man-made cavities, in cliffs and buildings, and is often seen around the Mediterranean.

And lastly, here is *Apus apus*, the common swift, flying ten months of the year without cease. It breeds across a vast swathe of the world, from western Europe to China, wintering in Africa for eight months of the year. It is by far the most abundant of the swifts, meriting its vernacular name in global terms and listed as the eighth most numerous wild bird species in existence. Worldwide, there are estimated to be somewhere between 95 million and 165 million of them sailing across the skies.

The common swift is not quite as numerous as the

Distribution of the common
swift. Black indicates its
wintering range and dark
grey its breeding range.

red-billed quelea, a little sparrow-like bird that lives in
sub-Saharan Africa, whose post-breeding population
numbers around 1.5 billion, and significantly less, too, than
the now-extinct passenger pigeon whose population was
estimated at over 3 billion in the early nineteenth century.
Flocks of these pigeons once stretched for hundreds of
miles on migration, darkening the sky for hours. Their
rapid decline, brought about by ruthless individual and
commercial hunting, combined with destruction of forests,
led to their extinction. The last wild bird is thought to
have been shot in 1901, while one captive individual lived
on in a zoo until 1914. The fateful story of this bird is
an awful warning that still resonates today.

Still, the common swift is plentiful. Abundant enough
to touch the lives of billions of people around the world
– so long as we wake up and look after them.

I have experienced nearly all these birds simply as an
armchair naturalist, but it has been a revelatory journey
and definitely more exciting than exploring my own family
tree. Never have I found a record of a distant aunt raising
her family behind a waterfall, nor do I believe my forebears had such varied and interesting feet.

EARLY MAY

Swifts arrive late in the sequence of spring. It is the perfect time of year. In my garden the hawthorn blossom is opening, cowslips are like fistfuls of golden keys in the mini-meadow, while the white honesty that billowed with flowers for the whole of April, glowing bright at dusk, is starting to lose its petals, fraying into tattered lace. Frogs have appeared in the pond, where they bask on a round stone in the water, while others rest on the weed, half-submerged with heads above the surface, enormous protruding eyes rimmed with gold. There has been no frogspawn this year, but the pool seems to have become an amphibian lido, scores of frogs of different ages and sizes lounging about. The newts rise slowly to the surface of the water, idling in the warmth. They are warier of my presence than the frogs, vanishing into the weed unless I am stealthily slow and silent.

Front garden lilacs scent the streets. Sparrows, which also nest in the cracks and crevices of our homes, are feeding ever-hungry young, their insistent cheeping a constant. A family of young great tits calls from a nest box – *zizz zizz zizz*! All through the day the parent birds fly back and forth with beakfuls of caterpillars, snatching a few sunflower seeds from the feeder to keep themselves going. I am lured into the garden whenever I have a spare minute, and though I may not be aware of it, at some level I am listening for swifts.

Out of the blue on the second day of May they appear, a squadron of four, screeching low overhead. 'Swifts!' I shout, though there is no one around to hear. These are the early ones, harbingers of the main arrival. The next day I do not see a single one. That does not necessarily mean they are not here. Newly arrived birds tend to be quiet for several days, feeding up and renewing their energy after the long journey from Africa.

The following day I go off to find butterflies. A few miles out of town is an old limestone quarry, mined on and off for the last 2,000 years for copper, silver and, most of all, the rock itself, an essential ingredient in the industrial blast furnaces of Coalbrookdale during the nineteenth century. These quarries finally closed in the 1930s and nature has reclaimed the ground, with woodland, sunny glades and a flourishing array of wild-flowers and butterflies against a dramatic backdrop of cliffs, much favoured by rock climbers and the occasional peregrine.

I am hoping to see the pearl-bordered fritillary, a butterfly whose flight season is almost over when the swifts arrive. For most of their brief lives they exist as

caterpillars, munching their way through violet leaves
before disappearing into a chrysalis like a dead leaf, sometimes slung low on an old bracken stalk, from which they
miraculously emerge just three weeks later. Their tawny-
orange wings have an exquisite tracery of black lines, and
on the underside a border of tiny white pentagons – not
unlike the home symbol on a mobile phone.

The flight of this fritillary could scarcely be more
different than that of the swift. In its entire life it may
fly little more than half a mile, staying low over the ground,
alighting on dandelions for nectar, the female seeking out
violet plants on which to lay her eggs. They live in colonies and are highly vulnerable to changing habitats; if the
violets disappear, so do they. When people made a living
from our ancient woods, coppicing and charcoal-making,
there were glades, open rides and sunlight, so flowers grew
abundantly. When such traditional rural ways of life disappeared, our woods grew darker and populations of this
once common butterfly dwindled too; it is now listed as
a threatened species.

It has been on the wing quite early this year, reaffirming
its old name, the April fritillary. Perhaps I am too late, or
maybe they are hiding in the vegetation from the gusty
breeze, but I do not see a single one. My search is rewarded,
though, by a pair of shimmering green hairstreaks, jewel-
like in the bracken, and a couple of woefully named dingy
skippers, intricately patterned in brown and cream. Then,
at last, I look up and there against a cloudless sky half a
dozen swifts are racing overhead. They could be on passage
to their breeding places or they might already have reached

their destination: several pairs nest in cracks in the cliffs here, as they may have done for thousands of years.

That evening I walk into town just before sunset. It is fine and swift calls are ringing over the streets, a wave of just-arrived birds feeding and circling around the houses. The Portland Bird Observatory in Dorset has recorded the first big arrival of swifts today, my birds perhaps part of this influx. Their instinct guides them straight back to their nesting holes. As the light ebbs, one of them slips under the eaves of a building I am watching, another posts itself into the stonework further down the street. The sky is now a dusky blue and, against the waxing, gibbous moon, a late swift flies, a fleeting silhouette on silver.

The swifts arriving in early May are generally those that are ready to breed. They are likely to be at least three or four years old, some of them last year's wild-child birds that raced and screamed across the skies, seemingly for sheer joy. Their first summers were purposeful though: their time was spent finding a mate and, crucially, a nest hole.

Ten months on and they are back, guided by a memory map to their claimed nesting sites, fast as an arrow. Occasionally, their recollection is slightly wrong; some swifts have to make several attempts to get their flight path on course for the nest hole. Friends tells me about the swifts that nest under their eaves, one of the pair particularly prone to misjudging the position of the entrance, quite often flying through an open first-floor window, then shuffling around on the floor until someone rescues it and takes it back to the open window.

I have watched video footage of these just-arrived birds and seen how they head straightaway to the nest cup at the back of the box, made during their last visit, a slightly ramshackle creation of feathers, tree flower sepals and other wind-blown booty, stuck together with saliva. There is a sense of homecoming as they arrive and a feeling of purpose.

The pair seldom arrive on the same day, but if this does happen it is likely to be coincidence; their bond exists only for the breeding period, and outside this time they live separate lives. When the second bird returns, usually within a few days, their initial greeting is sometimes hostile. They raise their wings, scream at each other and lock feet as they do in fights. Gradually, though, they accept and start preening one another, settling amicably on the nest side by side.

In the days when a bird is alone, before its mate returns, it may not always stay in its nest hole overnight. The all-seeing eye of the webcam shows one restless and ill at ease, first roosting vertically on the back of the box, then moving to the nest, then back again, and finally leaving, flying into darkness shortly before midnight. For a lone swift, the sky may be a better place to sleep, among its kind. Others, though, may be quite content to roost by themselves; these birds are all individuals.

One of the earliest and most vividly descriptive accounts of swifts sleeping on the wing was written in the late nineteenth century. Two brothers, Aubrey and Cyril

Edwards, spent many hours watching swifts at twilight in 1886 and published their observations in *Nature*, the illustrated journal of science, and again in 1914 in the proceedings of the Dorset Natural History and Antiquarian Field Club:

> If you will watch the Swifts at sunset on a fine evening you will see them all gather together and fly about in all directions, like distracted spirits, for some time. Then, as the dusk creeps on, you will see them get into order, form themselves into a flock, and ascend into the sky in wide spirals, screaming all the time. They will disappear from sight several times, but come round again, and at last they will rise so high that they are lost to the sight of the unaided eye, though with a binocular you can see them for some minutes longer. Then the sound ceases, and the stars are out.

If after watching them rise up you had sat on a tomb-
stone under the eaves where they build, till half-past ten
(with watchers on the other side of the church) to make
sure that no bird returned to the nests, and on other
nights alone till eleven, you would know each time that
they didn't come back to their nests that night.

Sightings by a French airman during World War I added
further weight to the theory:

As we came to about 10,000 feet, gliding in close spirals
with a light wind against us, and with a full moon, we
suddenly found ourselves among a strange flight of birds
which seemed to be motionless, or at least showed no
noticeable reaction. They were widely scattered and only
a few yards below the aircraft, showing up against a
white sea of cloud underneath. None was visible above
us. We were soon in the middle of the flock, in two
instances birds were caught and on the following day I
found one of them in the machine. It was an adult swift.

The poet Wilfred Owen was also familiar with the idea
of swifts roosting on the wing when he wrote and rewrote
his 'Ode to the Swift', between 1912 and 1917:

But when eve shines lowly,
And the light is thinned,
And the moon slides slowly
Down the far-off wind,

Oh then to be of all the birds the Swift!
......
And there thou sleepest all the luminous night,
Aloft this hurry and this hunger,
Floating with years that knew thee younger,
Without this nest to feed, this death to fight

The very romanticism of the idea that swifts might sleep during flight may have been an obstacle to its acceptance by scientists, some of whom remained deeply sceptical. As late as 1952, the *Handbook of British Birds* stated that it was improbable that birds would sleep on the wing. It was not until the Swiss ornithologist Emil Weitnauer boarded a little plane to search the skies at dusk and dawn that the theory began to gain acceptance. He too found parties of swifts circling high.

Two other Swiss observers witnessed swifts while watching the moon through a telescope. And clinching evidence was provided by radar tracking. Marconi recorded swifts on film over southern England and the English Channel, and Derek Bromhall described the astonishing results in *Devil Birds*: 'As the sun set and darkness fell, all detail on the screen was blotted out as thousands upon thousands of swifts rose into the night sky, in what has come to be known as their Vespers flight.'

All non-breeding common swifts are now known to take part in night ascents. About twenty minutes after sunset they spiral into the sky, their wings quivering. They may rise as high as 10,000 feet, seeking out warmer layers of air to pass the night. Here they will alternately glide and beat their wings – slightly more slowly than during the day – to maintain height.

During their nocturnal flights, swifts orient themselves into the wind to prevent themselves drifting too much. Using radar tracking and altimeters pasted to the backs of swifts, scientists in Sweden were surprised to find that in strong winds, the birds neither increased their flight speed nor sought out higher altitudes where the wind was gentler, even though this meant that they were driven backwards – perhaps sixty miles or so. Staying in a warm but windy aerial zone saves them more energy than spending time in a less blowy but colder atmosphere, even though they must make up the distance the next day. For a swift such a journey is not a big deal. It will have the chance to forage as it flies, and perhaps a tailwind too.

The majority of their nights are spent at between 3,000 and 7,000 feet; only on warmer nights do they ascend beyond this. In particularly poor weather they will simply fly elsewhere.

It is suspected that they have the ability to sleep with just one half of the brain while the other half stays alert. Other birds, such as mallard, do this – it is how they manage to sleep with one eye open, watching out for predators. In fact, evidence is emerging that many bird species have the ability to take micro-naps as they fly, enabling them to keep going on their epic migratory journeys across the globe.

Recent radar research carried out in the Netherlands shows that swifts ascend high into the sky at twilight, not only at dusk but also at dawn. This discovery throws doubt onto the purpose of these flights; they may not be simply for sleep. It has been suggested that since the magnetic compass of birds is calibrated by twilight cues, swifts may use these ascents to assist orientation.

Even now, their aerial habits hold many mysteries.

6

MAN-MADE HOLES AND MYTHS

Swifts lived in a humanless world for millions of years. During this time, *Apus apus* would have nested in fissures in cliffs and holes in trees within the wildwood that once swathed much of the globe, just as many within the Apodidae family still do. Today, in the few remaining primeval woods of central Europe, several hundred pairs of common swifts raise their broods in woodpecker-drilled holes in trees. In Poland they largely nest in hornbeams; in Germany they choose oaks. In Lapland, a small population breeds in the dwindling forests, and a handful still nest in dead Scots pine trees in the remnant Caledonian forests of Scotland.

Alternative nesting opportunities emerged as the human population expanded and woodland clearance gradually altered the landscape. Common swifts discovered holes in the houses people built for themselves, man-made vertical edifices that had similarities with cliffs and provided an abundance of dark crevices, ideal for their needs.

Early human dwellings were often single storey, but as building techniques became more sophisticated and ambitious, they got bigger. The Romans gave swifts a massive boost through the construction of aqueducts, temples and public buildings, and the beauty of these was their height. This can make the difference between life and death to fledgling swifts taking their first flight from the nest. Durability was another feature of their work: many relict buildings from the Greco-Roman world survive today. Some 2,000 years after its construction, the magnificent Pont du Gard, near Nîmes in France, the tallest aqueduct bridge built by the Romans, continues to attract breeding Alpine swifts each summer.

The Pont du Gard, near Nîmes.

One of the most distinctive inventions of the Romans was the convex terracotta tile, a design that has been used across the Mediterranean ever since. These glowing roofing materials look wonderful and, by sheer good fortune, also make perfect nesting sites for swifts, sparrows, starlings, bats and geckos.

Later on, rocks were hewn and hoisted to build cathedrals, churches and castles, bringing cliff-like structures to an array of places. Cracks and cavities appeared as stone and mortar weathered and aged. Towers with ventilation shafts provided sheltered space for whole colonies of swifts to nest.

Swifts exploited perishable roofing materials too: thatched cottages offered snug nesting holes, as Gilbert White, curate, naturalist and author of *The Natural History and Antiquities of Selborne* (1789), observed. He noted their preference for 'crannies of castles, towers and steeples', but also spotted them nesting in the cliffs of the local lime pit and in the thatch of 'the lowest and meanest cottages'. Here they were vulnerable to cats and people, who beat them down 'with poles and cudgels as they stoop to go under the eaves'.

Curiously, though, it was the industrialisation of the nineteenth century that gave them their greatest chance to increase in abundance. Cities expanded and warehouses and factories sprang up, built with bricks, which, just like stone, soon proved friable enough to accommodate wildlife.

Weathering is a great thing; a wearing down of materials through rain, sunlight and freezing temperatures that slowly erodes their crafted integrity and lets nature move in.

Humans have lived within inches of swifts for thousands of summers, but we are often unaware of our strange attic companions. Few can fail to notice the presence of house martins or swallows; their nest-building goes on in full view. Flying back and forth adding little lumps of mud, we hear their rapid, babbling chatter through open windows, and see chicks peep out of their nests, jostling for the best position to be fed, an openly amiable addition to the household.

Swifts, though, are elusive, vanishing under the eaves in the blink of an eye. A parent bird makes fewer trips to its young, collecting hundreds of insects for each meal, glued with saliva into a bolus, which is held in its enormous throat and delivered once every hour or so, very fast. *Vroom.* You hear a rush of wing-beaten air and, if you are quick, catch sight of the bird darting into a crack with a rustle of closing wings. The soft, piping calls of the young birds are seldom heard; only when the parent bird arrives with food do you catch their silvery sounds, a whispered version of the adult's cry. Their lives remain largely hidden; they are so close to us, yet so little known.

This elusiveness may have contributed to their reputation for evil in folklore. Quick as Lucifer, a swift vanishes, glimpsed like a spirit from another world. While it has generally been considered lucky to have swallows or martins nesting in your house, swifts have at times been regarded with suspicion. In Kent, a farmer from the nineteenth century is on record saying, 'Knock them black swifts down sir, they are regular limbs of the devil' and referring to them as 'black imps'. Country names such as devling, devilet, sker-devil and screech devil have perpetuated the myth.

People have wondered forever about the lives of birds which arrive in early spring and depart in late summer, and the theory that they might go somewhere warmer for the winter has been debated for thousands of years. Sailors told stories of swallows perched in the rigging of ships at sea, which leant supporting evidence to the idea, but hardly conclusive proof. In the fourth century BCE, Aristotle wrote that cranes migrate from one end of the world to the other and suggested that swallows and kites did too. However, he also gave space to another speculative theory, suggesting that a great number of birds simply go into hiding over the winter. Swallows, he related, had often been found in holes in the ground, quite denuded of their feathers.

Swifts, swallows and martins have been muddled up for as long as people have looked up into the skies, and it is likely that all these summer visitors were believed to share the same wintering habits.

For centuries, Aristotle remained an influential source for writers describing the natural world, and both theories continued to be propounded in the absence of hard scientific knowledge. So it was that people continued to claim that birds slept away the winter in ruined buildings, towers, caves or hollow trees. Most preposterous of all, by the sixteenth century it was suggested that swallows disappeared into lakes and ponds as the days shortened and temperatures dropped, hibernating in the mud at the bottom.

This notion, described anecdotally, was given weight by Olaus Magnus, Catholic Bishop of Uppsala, in his *History of the Northern Peoples* published in 1555, and first related in English by Richard Carew in his Survey of Cornwall in 1602:

> He saith, that in the North parts of the world, as summer weareth out, they clap mouth to mouth, wing to wing, and legge in legge, and so after a sweete singing, fall down into certain great lakes or pooles amongst the canes, from whence at the next spring, they receive a new resurrection; and he addeth for proof thereof, that the Fishermen, who make holes in the ice, to dip up such fish with their nets, as resort hither for breathing, doe sometimes light on these Swallowse, congealed in clods of slymie substance, and that carrying them home to their Stoves, the warmth restoreth them to life and flight!

Magnus, who wrote his epic work in Rome, where he lived in exile after the reformation of the Church in Sweden, also told the sailors' story of the great Norway sea serpent, a 200-foot monster that coiled itself around ships and devoured their crew. Such myths were constantly repeated by fishermen and sailors, with many variations as to the precise nature of the beast. These stories, which gained much popular credibility, were likely based on sightings of real marine creatures, such as the lamprey-like oar fish, giant squid, eels and porpoises. Imagination laced with fear magnified and distorted them into the monsters of folklore. Even today, sightings of 'sea serpents' continue to emerge, flashed around the internet on mainstream media.

Magnus' book would be widely translated and respected, though the more fanciful sections would later attract scorn and condescension – he was writing, his critics wrote disparagingly, in the Dark Ages. Recent commentators have been more generous, valuing his history of vanished times and its rich folklore. Meanwhile, the *Carta Marina*, his map of the North Atlantic seas, on which the sea serpents are illustrated, is now acclaimed as the earliest-known depiction of the Iceland-Faroe front, enormous swirls of water where the Gulf Stream meets cold waters coming down from the Arctic. These eddies are depicted on the *Carta Marina* with perfect accuracy.

Two centuries later, his fellow countryman Carl Linnaeus, the pioneering ecologist and inventor of the modern system of binomial taxonomy, still firmly believed that swallows spent their winters in hibernation. As professor of botany at Uppsala University, his knowledge and skill were highly renowned, but his grasp of bird-related matters was a little shaky. The English naturalist Thomas Pennant levelled scathing criticism at him, writing: 'As to ornithology he is too superficial to be thought of.' Meanwhile, another English scientist speculated that swallows hibernated on the moon, which at that time was believed to be vegetated and habitable in a similar way to Earth. Such alternative explanations would continue to surface until conclusive evidence was discovered to support the theory of migration.

The enduring power of the hibernation theory appealed to people's 'common sense'. Mammals such as hedgehogs and dormice were known to survive the winter through hibernation – why should the same not be true of birds? The concept of small, just-fledged birds, flying vast distances across the world to find warmth and food was a hugely challenging one. According to traditional Christian orthodoxy, humans were the pinnacle of God's creation. Navigation still presented often mortal difficulties to experienced sailors and explorers; surely birds could not have an intelligence that people lacked?

One of the first people to describe the habits of swifts in detail was Gilbert White, the parson naturalist of Selborne who recorded his observations in a series of letters published in 1789. He noted their nesting habits, method of feeding their young with a ball of insects, arrival and departure dates and their annual squabbles with sparrows. Inevitably, he was drawn into the debate about where they and their fellow birds, the swallows and martins, spent their winters. Into this mire of doubt and uncertainty he waded, pulled in both directions. He accepted that the majority of house martins, swallows and swifts flew south to hotter regions, describing with vivid detail a scene he had witnessed in late September:

If ever I saw anything like actual migration it was last Michaelmas-day. I was travelling and out early in the

Gilbert White near his home in Selborne, Hampshire.

morning: at first there was vast fog; but by the time that I was got seven or eight miles from home towards the coast, the sun broke out into a delicate warm day. We were then on a large heath, or common, and I could discern, as the mist began to break away, great numbers of swallows clustering on the stunted shrubs and bushes, as if they had been there all night. As soon as the air became clear and pleasant, they all were on the wing at once; and, by a placid and easy flight, proceeded on southward towards the sea.

His reflections on looking at recently hatched nestling swifts also reveal him thinking about migration: '. . . while we contemplated their naked bodies, their unwieldy, disproportionate abdomina, and their heads too heavy for their necks to support, we could not but wonder when we reflected that these shiftless beings, in a little more than a fortnight, would be able to dash through the air almost with the inconceivable swiftness of a meteor, and, perhaps, in their emigration, must traverse vast continents and oceans as distant as the equator'.

Nonetheless, he had reservations about the ability of late-fledged broods to achieve this. Hibernation seemed to him a more likely option: 'Did these small weak birds, some of which were nestlings twelve days ago, shift their quarters at this late season of the year to the other side of the northern tropic? Or rather, is it not more probable that the next church, ruin, chalk-cliff, steep covert, or perhaps sand-bank, lake or pool . . . may become their hibernaculum and afford them a ready and obvious retreat?'

Bats were known to wake from their torpid state to take advantage of warm spells of weather in the winter, and White believed hirundines did the same thing. Observing house martins feasting on insects over nearby fields in early November, a full month after their vanishing in early October, he concluded that:

> . . . many individuals do never leave this island at all, but partake of the same benumbed state; for we cannot suppose that, after a month's absence, house martins can return from southern regions to appear for one morning in November. From this incident, and from

repeated accounts which I meet with, I am more and more induced to believe that many of the swallow kind do not depart from this island, but lay themselves up in holes and caverns, and do, insect-like and bat-like, come forth at mild times, and then retire again to their latebrae, or lurking places.

His ambivalence about their wintering habits is expressed in a letter written in 1794. He notes that the plumage of swifts on arrival in spring is glossy and dark, but by the time of their departure they are somewhat 'weather-beaten and bleached'. Wondering about this, he asks: 'Now, if they persue the sun into lower latitudes, as some suppose, in order to enjoy a perpetual summer, why do they not return bleached? Do they not rather, perhaps retire to rest for a season, and at that juncture moult and change their feathers, since all other birds are known to moult soon after the season of breeding?'

He also relates a sorry story told to him by 'a clergyman, of an inquisitive turn', who assured him: '. . . when he was a great boy, some workmen in pulling down the battlements of a church tower early in the spring, found two or three swifts among the rubbish, which were at first appearance dead; but on being carried towards the fire, revived. He told me that out of his great care to preserve them, he put them in a paper bag, and hung them by the kitchen fire, where they were suffocated.'

White had hoped to settle this great mystery of natural philosophy beyond doubt, but his arguments often suggest he was actually looking for confirmation of the hibernation theory. He seems inclined to favour this explanation for later-fledged birds, doubting how young, inexperienced birds could undertake such a long, arduous journey, while simultaneously accepting the probable migration of earlier fledgers. It was a hugely complex issue both in terms of scientific enquiry and mental boundaries. For humans such journeys would be a huge challenge; how much more so for a fragile bird?

Thirty years later opinion was starting to change. Henry Reeve, a Suffolk physician of eclectic interests, wrote an essay published in 1809 on the torpidity of animals and expressed exasperation with those who continued to support the theory that swallows might hibernate: 'There is scarcely a treatise on ornithology which does not allude to the submersion of swallows during the winter, as a fact almost as well known as their flying in the air during the summer . . . And what is the evidence in favour of so strange and monstrous a supposition? Nothing but the most vague testimonies and histories repugnant to reason and experience.'

Perhaps it was just too good a story to die. Like tales of Scotland's Loch Ness monster, this mystery fascinated people; they wanted to believe it was true. The myth also contained a wisp of truth: swift nestlings have the ability to go torpid in cold weather, enabling them to conserve energy when food supplies are scarce. But when they fledge the young swift loses this capacity, and adult swifts simply keep flying and hunting for food to sustain themselves through bad weather.

Within the scientific world, it was Edward Jenner's observations and arguments on bird migration, published posthumously by his nephew in a paper read to the Royal Society in 1823, that finally laid the ancient myth to rest. Jenner, the scientist who invented a vaccination against smallpox, was deeply curious about the natural world and undertook detailed observations of birds and other wildlife. Hibernation intrigued him, but he cast scorn on the notion that summer-visiting songbirds and hirundines – among which swifts were at this time classified – might overwinter in river banks and ponds: 'At such a time, what can be the inducement to them and their young ones, which have but just began to enjoy the motion of their wings, and play among the sun-beams, to take this dreary plunge? I have taken a swift about the 10th of August, which may be considered as the eve of its departure, and plunged it into water but like the generality of animals which respire atmospheric air, it was dead in two minutes.'

He also noted that hibernating creatures such as the hedgehog emerged from their winter sleep in poor condition: '. . . fat is its only source of nutrition for the winter, which, by the time the sun rouses it to fresh life and activity, is exhausted, and the animal comes forth thin and emaciated . . . If on the first day of its appearance, a martin, a swift or a redstart be examined, it will be found as plump and fleshy as at any season during its stay.'

The truth of their journey is perhaps even more marvellous than the myths. Bird migration is a wondrous, mind-boggling feat, stretching our imaginations even in the twenty-first century, when we know it is real, with evidence provided by ringing records and more recently by a wealth of new data uncovered through the use of light-logging geolocator tags. Even today, though, the mechanics of how birds navigate are not fully understood. It is little wonder that White and others were drawn towards the familiar explanation of hibernation.

Curiously, bird-watchers have recently reported seeing swifts in December. Not hibernating, but flying. Reports of swifts were recorded in the last month of 2015 in Northern Ireland, Norfolk, Kent, Yorkshire, Sussex and Dumfriesshire. There were twenty November records from 2000–2015. Is this a result of warmer winters? Possibly, but the absence of records of swifts in January and February suggests that these stay-behind birds struggle to survive.

7

FLIGHT AND FEET

In Gustave Doré's wood engravings of angels, produced for John Milton's *Paradise Lost*, Satan is depicted with the leathery wings of a dragon, while all the others have the feathered wings of birds. Western artists generally illustrated angels with white plumaged wings but Doré's are printed in black ink. Dramatically long, they are redolent of swifts. Angel wings in some Byzantine icons are similar: jet black, tapered and narrow. I find myself wondering how these angels could get airborne with their weighty-looking human bodies. A foolish question, of course. These illustrations are artistic representations of spiritual beings, their wings merely symbolic of their ethereal nature.

For a flesh and blood bird, though, aerodynamics are everything. Honed through millions of years of evolution, their bodies, bones and feathers are fine-tuned to an aerial life. Bird skeletons have a fragile appearance but they are tougher than they look. Mostly hollow, their bones are nonetheless higher density than those of mammals of a similar

Gustave Doré engraving for
John Milton's *Paradise Lost*, 1866.

size. Inside, they are criss-crossed with struts, giving them the strength needed to withstand the pressures of flight.

Highly agile, swifts can adjust the shape of their wings according to the myriad requirements of life in the air. They continually change the spread and shape of their wings to maximise speed and minimise energy expenditure. At Wageningen University in the Netherlands, scientists set up a wind tunnel experiment using pairs of swift wings collected from dead birds, to demonstrate the physics of their shape-shifting wings. They showed how the sweep of the bird's wings affected flight or sink speed;

how extending them suited slow glides and turns; and how swept wings were superior for fast ones. At Lund University in Sweden, Anders Hedenström leads a team of scientists exploring the intricacies of avian flight mechanics, and they have carried out trials of live, just-fledged swifts in wind tunnels, releasing the birds into the air after a few days.

The wing skeleton of a bird has a close similarity with that of the human arm. Like us they have a humerus, radius and ulna, wrist, hand and digits, though these are fused. Avian bones, however, clearly serve a different purpose to human ones and have evolved accordingly. Each bird species, too, will show adaptations in structure, necessary modifications for the varied requirements of its life.

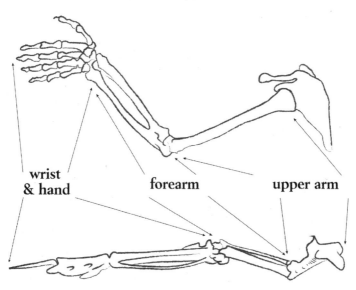

wrist & hand   forearm   upper arm

Swift wing / human arm.

The bones of the hand in most birds are fused rigidly together, meaning that the ten primary feathers attached to the hand operate like a Japanese fan, all moving at once. However, the swift has flexibility at the knuckle joints, giving it greater mobility. A bird with such long wings might not easily be able to tilt and swerve, but this adaptation enables it to close the five outer primary feathers growing from its finger bones on one wing, while keeping all ten primaries on the other extended. This imbalance causes it to flip towards a vertical plane and turn with ease, while the head remains horizontal throughout.

With dramatic humour, Lyndon Kearsley, an ornithologist living in Belgium, describes the stocky, shortened humerus – the inner wing bone – as the feature that stops both swifts and hummingbirds from having their arms ripped off. Or, as he then puts it more soberly, the strength of this bone allows both of these closely related families to beat their wings very quickly and strongly.

Lyndon has been ringing birds for over fifty years and tells me that when you hold a hummingbird in your hand you can sense its likeness to a swift. In both birds the primary feathers make up most of the wing. They are stiff, with strong central feather vanes that ensure little air passes through, making them tough as oars. 'The hummingbird is, one could say, just a swift that discovered sugar, making the leap from insectivorous to nectar-eating and splitting from them very early in their evolution,' suggests Lyndon. The stocky humerus gives hummingbirds the strength to beat their wings fifty times per second as they hover over nectar-filled flowers and enables swifts to endure their constant aerial existence.

There is a slight roughness to the swift's wings, the result of overlapping feathers, which enhances its aero-dynamic performance. Flapping causes turbulence along the outer two-thirds of the bird's wings and this spirals along the top of the leading edge all the way from the wrist to the tip of the wing and beyond. It is like a virtual wing, lengthening the surface area beyond the actual wing, over which the air above flows smoothly, giving efficient glide performance. 'The swift's flight adaptations have an elegant simplicity,' says Lyndon. Their aerial technique and instinct for exploiting favourable winds and air currents is unsurpassed, using the least amount of energy for the best result.

The swift's scimitar-shaped wings taper towards the tip like a falcon's. This reduces drag, allowing them to fly faster, the length enabling them to glide and take advantage of air currents, such as gusts of wind driven up as it buffets buildings or rises in thermals over the sea. Wider wings, such as those of an eagle, are even better for gliding, with a large surface area to catch warm air currents rising from the ground, but such wings are heavy and unsuited for fast flight. To retain height and motion the swift beats its wings, alternating fast strokes with gliding. Watch them and you see how they flicker in rapid movement then stretch out, fractionally down-curved, to drift along. Slightly forked, the tail is at once a rudder and a brake, helping it to change direction and slow its speed as required.

An airman in World War I reckoned from his speedometer that swifts flying close to his plane were moving at 68 miles an hour. More recently, scientists in Sweden have measured their speed in wind tunnels, recording one bird at 69.3 miles per hour. This evidence confirmed that swifts are the fastest birds in level flight. Incredibly agile, they can even maintain their racing speed while flying upwards. Young swifts in screaming parties are the speediest, hurtling through the sky in packs, skimming roof tops, scything the air. The dive of the peregrine falcon may be faster – it has been recorded hurtling downwards at 200 miles per hour – but its speed is greatly enhanced by gravity.

Within the last few decades, ornithologists in Switzerland and Germany have observed no fewer than nineteen distinctive variations of swift flight, learning the swift's body language through careful observation. According to its purpose, it flies at varying speeds using a range of techniques: turning sideways to take corners while hunting; a shallow glide as it sheers across the surface of a lake, scooping up water to drink, shivering as it rises again to shake off drops of water from its feathers; alternate gliding and soaring while asleep. They have noted the flight of parent birds scouring the countryside for insects, zig-zagging to confuse a hunting hobby; witnessed swifts remedying a small problem of ruffled feathers with a short fall, whirling like a dead leaf in the air; courtship flying; the quivering Vespers ascension flight. Watched closely, these birds reveal an array of aerial skills to deal with different situations. Swifts are renowned for speed and screeching but their flight is often leisurely and silent.

Scanning the air for food, they are focused only on finding insects.

The vast distances travelled by a swift during its lifetime are sometimes described in terms of journeys to the moon and back – 477,710 miles. We can only generalise – each swift is an individual – but a rough calculation of distance travelled can be made by multiplying their average speed by time spent in the air. A breeding bird is in flight for roughly ten and a half months each year – 7,500 hours. The speed at which it flies varies, but at an average of 30 mph it would cover 225,000 miles per year. At a speedier 45 mph it would fly 337,500 miles. Non-breeding birds never stop at all, so the annual distance for the first three years of its life would be even greater. The average life of a swift is around seven years, so at 40 mph it may fly over two million miles – approximately four trips to the moon and back.

Flight is not simply a means of moving across the world, a way to catch prey and feathers for a nest. It is a whole language through which these birds make their intentions known to each other, the very essence of the swift's being. Looking up, we rarely pause to give these birds more than a passing glance. When we do, we are amazed by their verve and speed but seldom look long enough to understand much more. Fortunately, there are people who spend entire summers watching swifts so they can learn and unravel their codes of behaviour. Their dedication is inspirational; by watching more frequently and for longer, we can start to tune into the swifts' behaviour and read their flight language.

In a gathering of a dozen or so, I try to follow an individual bird as it arcs overhead, etching the air with invisible traces. An architect and photographer, Laurent Godel, has found a way of revealing them. In a series of multiple exposure, high-speed shots over Barcelona, he shows the very wing-beats of birds sweeping across the sky, flying circuits over the city roofs and around buildings, their flight trails ravelled like loosely coiled barbed wire. He catches them over the undulating roof and walls of La Pedrera, a building designed by Antoni Gaudi early in the twentieth century, their circling flight trails mirroring the curves of the building.

Another picture shows a single swift streaking past the Columbus Monument, while several more race in a vortex of wings around this ornate edifice. Four winged figures symbolise the fame of Spain, carried across the world, but in this image it is the supreme aerial skill of the swift and its own global abilities that leap out at the viewer. Look closely and you can see every downstroke, upstroke and glide. 'We are often too busy to really look at the world we live in. Taking pictures allows me to stop for a moment, focus on my surroundings, and uncover a bit of the magic that is present everywhere,' says Godel.

Watching them circle overhead on a fine evening, I am reminded of something. When I was a child, one particularly cold winter, my mother took us skating on a nearby lake, an old mill pond about half a mile long. This may sound shocking to young ears, reared in times of greater caution, but the lake was relatively shallow – so she told us – and temperatures had been below freezing for a week or more. We had a box of ancient skates and I rootled out a pair that fitted pretty well, padded out with some thick woollen socks. I had never skated before, so a chair was brought for me to push along the ice while my sisters raced away. Slowly I got the hang of it: push off with one foot, glide with the other, bring your feet together; push off with the other and glide. Then I shoved the chair away and flew across the lake – as near, that is, as an earthbound creature can. I am not sure now how competent I really became that day, but in my probably faulty memory I sped around the ice, sweeping across in vast circles under a white-grey sky. Just like the swifts above me now in the blue of summer.

I wonder about the common swift's Latin name, *Apus apus*, meaning without feet. Carl Linnaeus, who gave them this binomial, was curiously persistent in his belief in the hibernation myth but he is unlikely to have meant the name *Apus* to be taken literally. The name pre-dates him by over a thousand years, Aristotle referring to the footless bird in the fourth century BCE. Perhaps it is simply a metaphor for the swift's aerial existence. Apus is also the name of a small constellation in the southern sky, representing a bird of paradise – also said to be without feet. This was first depicted on a celestial globe in 1598.

Seldom seen because swifts do not perch, their feet are tucked close to the body in flight. Observation of birds in nest boxes has made their legs and feet much better known. Their footwork is dire; walking is absolutely not one of the swift's strengths. Watch one on a webcam recording as it hobbles over the floor and you would be forgiven for thinking it was injured. Their legs are short but the feet are not small, just strangely shaped. When you see their long, curled-over claws you understand why they walk so awkwardly. Only a minimal proportion of the swift's life is spent on horizontal surfaces and their feet are just adequate for this within the safety of their nest holes.

If a swift falls, perhaps as a result of a fight in the nest, or is knocked down by torrential rain, or when a fledgling fails to get airborne on its first flight, it is not easy for it to take off. On the ground its wings drag, especially if it finds itself in long grass. If it lands on a smooth surface, however, or shuffles along the ground until it finds one, it can flap its wings more freely and has a chance of getting away – unless injured or snatched by a cat.

The majority of birds have anisodactyl toes, meaning three pointing forwards and one backwards. Most swift species have these, but the toes of the Apus genus are different. Those of the common swift are set in two pairs at an angle to each other. This arrangement allows them to hang vertically. Powerful, long and sharp, their claws give them an excellent grip, so they can roost on walls – as they very occasionally do – or on the sides of the nesting hole. They are also just the thing for fighting with, enabling the swift to lock onto its opponent's feet. Many a bird-ringer too has felt this strong grasp. Roosting in caves, chimneys or in trees is usual among other species of swift, though not all have this arrangement of toes; it is a feature of the Apodinae tribe.

Tree-roosting has rarely been observed in the UK but is quite regularly witnessed in Sweden, where swifts have been seen clinging to thin twigs or leaves on the outside of tree foliage, and also on a latticework mast in a vertical position. The Swedish ornithologist Jan Holmgren noticed how sometimes a swift might hang freely from a single leaf. 'Then, its position was more crooked, with the back almost downwards and with the wings pointing downwards.'

Such behaviour does not come as second nature. Holmgren noted that most of the birds had to make several fly-ins before succeeding in getting hold of a leaf or twig, sometimes making as many as twenty or thirty attempts. They would arrive ten to forty minutes after sunset, earlier in cloudy weather and later in clear weather, leaving very early in the morning if the weather was unfavourable, but sometimes staying until well after sunrise in good weather. Most of these observations occurred in August with the majority being newly fledged birds, recognisable by their pale faces and white-edged primary feathers. As in sky-roosting birds, these swifts collected into gatherings of a dozen or so up to a total of sixty. It is possible that tree-roosting occurs more commonly than previously thought. Very occasionally, *Apus apus* is forced out of its aerial realm by extreme weather and takes refuge on solid structures. As dusk fell on a May evening in 2012, during an unusually cold spell, common swifts were observed near Málaga in Spain, repeatedly trying to enter the narrow gaps in the windows of an apartment block, without success. Early next morning they were seen clustered vertically, like a feather boa, sheltering on a nearby wall. Huddled together, the birds stood a better chance of keeping warm. This phenomenon was also noticed in spring 2019, when swifts endured severe storms on their migration to Europe. Dozens of bedraggled birds were seen gripping the brick walls of houses.

Swifts sheltering after a storm, near Málaga.

It can happen anywhere on their journey, as described by the naturalist, Henry Douglas-Home. He recalled seeing swifts on migration in North Senegal: 'one night a violent sandstorm forced a pack of three hundred to cling closely to the mosquito wiring of our house'. That common swifts may, from time to time, stop for an hour or two during the night during migration and wintering periods has been confirmed by Swedish scientists. Anders Hedenström and his team at Lund University fitted tiny microdata loggers, each with an accelerometer and light sensor, to a number of swifts, capturing their journeys and flight activity over a two-year period. The devices were recovered from eleven birds in 2014 and eight the following year, providing a unique record of each bird's travels. A strong theme of near-constant activity emerged, with minor variations. One bird settled for four whole nights between September and April in its first winter, but stopped for just two hours over the same time period the following year. Several others ceased flying for up to two hours occasionally, while four of them were airborne the entire time. Roosting can be a life-saving strategy during bad weather and this is likely to be the reason why they periodically seek shelter from the sky. Thunderstorms, hail, intense rainfall and high winds pose a serious threat to swifts.

Their usual strategy for dealing with such conditions is to head off elsewhere, flying into the wind around depressions. During the summer, when the weather turns bad, non-breeders may gather in huge flocks, flying hundreds of miles to find better conditions, sometimes staying away from their colonies for up to a week. In the worst years, these young adults may return to Africa early, ensuring the survival of their generation. On the European continent, swifts fleeing bad weather sometimes take refuge over faraway river valleys and lakes. In 2009 a flock of at least 48,000 was counted above the shores of Lake Geneva, feeding on the plentiful insects.

Occasionally, swifts are utterly defeated by extreme weather, unable to fly away from it in time. A catastrophic phenomenon occurred in 2013, after a cold, rainy episode, when hundreds of swifts fell to the ground dead or dying in continental Europe. In Chevroux, on the shores of Neuchâtel in Switzerland, at the beginning of June, sixteen swift corpses were found on the pier, yellow-legged gulls gobbling them up. Many more were flying over the lake, hunting for insects, but with the temperature at 5°C, there were few to catch and the swifts appeared exhausted.

Meanwhile, a few days earlier at Limoux, in southern France, dozens of swifts were found dying on the ground in the town centre, some of them run over by cars. No one had ever seen anything like it. A long period of wet weather had, it seemed, caused the insects to stay at ground level among vegetation. With the skies empty of food, the swifts had starved.

Such desperate incidents are unusual. But extreme weather events are predicted to increase with climate change, perhaps bringing further pressures for swifts.

8

A TOWER AND
ITS SWIFTS

It would be more than 150 years after Gilbert White's *Natural History of Selborne* before any other account of the swift's life was published. David Lack's *Swifts in a Tower* appeared in 1956, telling the story of ten years' observations of a swift colony in the tower of Oxford University Museum of Natural History. Scientists and bird ringers have continued to monitor this colony ever since, making it the longest-running swift study in the world.

A wondrous Victorian Gothic edifice, the University museum has an extraordinary history. It was built as a place to display the anatomical and natural history collections previously housed in scattered locations around the city. Until its opening in 1860, the university had no centre for the study of natural sciences; indeed, some people considered it sacrilegious to subject God's creation to academic scrutiny and the scientists had to battle for it.

At that time, the boundaries between Christian faith and scientific exploration were less polarised, and religious arguments were used for and against its construction. Studying the natural world would reveal 'knowledge of the great material design of which the Supreme Master-Worker has made us a constituent part', argued the professor of medicine, Sir Henry Acland. Money was searched for and eventually found in the university chest, its financial treasury, which had benefited from the sale of bibles. Once again, there were objections, but Dr Pusey, a distinguished theologian and an advocate of high church doctrine, used his influence to persuade the university authorities.

A marvellous building took shape, embodying the natural world in its very construction, with cast-iron pillars supporting a glass roof, each one ornamented with leafy ironwork depicting such species as walnut, sycamore and palm. Stone columns, every one made from a different kind of rock, displayed British geology, while two talented Irish stonemasons, the O'Shea brothers, breathed life into the capitals and corbels with plants representing all the botanical orders.

As is common with the construction of large new buildings, the project ran over budget and funds for further carvings ran out. The O'Sheas offered to keep working without pay, but certain members of the congregation, the university governing body, accused them of defacing the building with unauthorised designs. These apparently included 'the unnecessary introduction of cats', carved around a window on an upper storey. Deeply upset, the

O'Sheas proceeded to carve parrots and owls over the main entrance, caricaturing the offending congregation members.

Scarcely had the building been completed than an intense debate on the theory of evolution took place within its walls at a meeting of the British Association for the Advancement of Science. Charles Darwin's *On the Origin of Species* had been published the previous year, in 1859, arousing controversy and hostility. The Bishop of Oxford, Samuel Wilberforce, who had supported the building of the science museum on the grounds that it would encourage the study of the wonders of God's creation, was outraged by the suggestion that man could be descended from apes.

He is reputed to have asked Thomas Huxley, who passionately defended Darwin's theory, whether it was through his grandfather or grandmother that he claimed descent from a monkey. The bishop's sarcastic derision backfired, with Huxley and the other pro-evolutionists garnering support from the audience.

The museum was extended in 1885 to house the ethnological collection of General Sir Pitt Rivers, a fabulous assemblage of magical objects, musical instruments, tools, masks, feather cloaks and ceremonial objects from around the world, each tagged with a tiny hand-written label.

While the hall soon became crowded with exhibits, fossils, rocks, stuffed animals, pinned insects and the sorry remains of the last dodo – a foot and a head – one part of the building remained empty. The tower had been built purely for ornamental reasons, rising sixty feet above the main roof, reminiscent of a French château. However, the room did not stay unoccupied for long. Pigeons found their way through the ventilation shafts and nested. Such messy creatures were not to be tolerated and it was decided that the shafts should be blocked at both ends to prevent their entry.

Some years later, workmen opened them up again in order to haul up slates to repair the roof. When the job was done the inner holes to the tower were once again blocked off but the exterior entries to the louvres were left open. No good for pigeons, but swifts took their chance and nested inside on the broken wooden ledges. This colony would become the focus of David Lack's studies.

Lack's path to this swift colony started early. A born bird-watcher, his childhood passion led him to study natural sciences at Cambridge, followed by a job as a biology master at Dartington Hall, an experimentally progressive school in Devon. Outside the classroom he carried out research into robins, colour-ringing each bird within his project area and watching their behaviour.

For four years he observed the activity of individual robins, and also carried out experiments with a scruffy old stuffed robin placed in their territory. During the breeding season it was fiercely attacked. Robins, he proved, do not sing because they are happy, nor to attract a mate. Song is their way of establishing and holding territory. The publication of *The Life of the Robin* would cause ripples of shock that our bright-eyed garden friend could be so fiercely territorial and combative. But the British fondness for the bird is unshakeable; robins simply do what they must to survive and thrive. In 1938 David Lack took a study trip to the Galapagos Islands, where he immersed himself in the evolutionary complexities of Darwin's finches, returning to England in 1939, after the outbreak of World War II. During this time he worked in radar development, which gave him unexpected possibilities to observe bird migration as well as enemy aircraft. Back home in 1945, he was appointed director of Oxford University's Edward Grey Institute. Founded eight years previously, its original remit had been economic ornithology, the study of birds' usefulness or detrimental effects in agriculture and horticulture. However, it switched its emphasis almost immediately to research the behaviour, ecology, evolution and conservation of birds.

While amateur naturalists such as Gilbert White had sought to learn about birds by observing their behaviour, this had not generally been the academic method of ornithology, which concerned itself largely with the study and anatomy of dead birds. Lack's research into living ones pioneered a new approach. His first project was to set up a long-term study of great tits at Wytham Woods, a 390-hectare woodland bequeathed to the university; it is a study that still runs to this day. Then, restless for a new project, he decided he would study swifts, too. Their awesome aerial life caught his imagination and their habits were largely unknown, making them a perfect subject for research.

When he first started looking around for somewhere to study swifts, he initially dismissed the museum tower

as impractical. Over 100 feet up and with the birds hidden
inside the ventilation shafts, it seemed to offer poor prospects for observation. What could he do? Gilbert White
had described in a letter written in 1774 how he had
removed roof tiles from a house where many pairs of
swifts nested, in order to investigate their habits, but such
intrusive action was clearly out of the question in the
twentieth century. A solution was found when Lack heard
about Emil Weitnauer, a schoolmaster in Switzerland with
a deep curiosity and love of swifts. Weitnauer had fitted
nest boxes under the eaves of his house, providing him
with opportunities to observe swifts closely. He would
watch them for over forty summers. Lack visited him and
returned with a plan.

The tower was soon fitted out as a research station,
with a platform four feet below the first row of ventilation
shafts. To reach it you first had to climb a steep, narrow
spiral staircase, part of it in total darkness, followed by a
perilous thirty-foot wooden ladder that would test the
nerves of scores of visitors in years to come. Another
platform was installed below the third row of shafts, and
ladders used within each room to reach the row above.
Next, wooden boxes were fitted inside, with lids so the
researchers could lift out the young for ringing and measuring. By April 1948, the work was done and, with some
trepidation, they waited for the birds to return. Fortunately,
all the birds accepted their modified homes and the experiment was underway. Sixteen pairs of swifts nested that
year.

The following year the wooden backs of the boxes were

replaced with glass, enabling Lack and the other swift observers to watch the birds from just a few inches away. Not only could they see these birds close-up, they could even see their breath condensing on the glass, a rarely experienced intimacy with a wild creature. 'It was fascinating to observe them in this way and to realise that we were watching behaviour which, except perhaps by our Swiss colleague, had not been seen by man before.'*

With typical post-war ingenuity, Lack and the team rigged up a special contraption for writing their notes: a section was cut away from an old petrol drum to turn it into a writing desk with a battery-powered light bulb and a writing board placed inside. 'With the open section of

---

* While Lack's research was the first formal, academic study of swifts, Weitnauer's experimental studies had been conducted since 1934. During the breeding season he too daily observed and recorded swifts' behaviour at the nest.

Earlier still, Henry Douglas-Home became fascinated by swifts when a schoolboy at Eton in the 1920s. During the evenings, before lights-out, he would lean out of his window and watch them in their wild chases – often while smoking a pipe 'borrowed' from his father. He managed to deflect the enquiries of the housemaster doing his rounds by attributing the smell of tobacco to the musty aroma of swifts.

Back at The Hirsel, Coldstream, the family home in Scotland, he set about making swift boxes, despite his father telling him it would be a waste of time: 'You'll never persuade a swift to nest in a nesting box. Nobody's ever done it and nobody ever will.' He constructed ten boxes, installing them just below the top-floor window sills. Like many swift-watchers, he did his best to prevent sparrows entering by fixing a movable flap over the holes, kept closed until the swifts arrived. He made many observations of the birds' behaviour by watching activity in the nest boxes, including the discovery of a nestling bird stone-cold and apparently dead. When the parent bird returned, he witnessed its recovery, 'warmed and cajoled into a full state of consciousness'. Douglas-Home would become a keen conservationist and pioneer of bird broadcasting for the BBC. Swifts continue to nest at The Hirsel.

the drum towards the body, and a notebook inside, we could sit facing the bird and could record in the light without the bird being able to see it,' wrote Lack. Meanwhile, some old black-out material was used to cover a window below the first platform, stopping light from shining through the gap around the ladder. Darkness was essential.

'Sitting in the darkness, very cold on a cold day and extremely hot on a hot one, we have not envied our fellow bird-watchers in their more conventional attitudes on marsh or mudflat. The passage of time is marked by the clock of Keble College over the way, always a little slow and a little out of tune, but otherwise we are cut off from the outside world as, undistracted, we seek to study, so far as a human being can, the way of life of the swifts.'

Swifts now came under scrutiny as never before. Lack spent long summer days in the tower watching them between 1946 and 1955, often accompanied by his field assistant, Elizabeth Silva. It was she, while watching alone, who first recorded the lengthy fights that take place between swifts when an interloper enters the nest hole. With feet interlocked, the birds wrestled for four and a half hours, the resident bird trying to push the invader out. She also observed the swifts' reaction to a cricket match between New Zealand and the university team on a nearby ground. The sounds of the crowd drifted upwards and roars of excitement could be heard as the visitors' wickets fell. When this happened, Elizabeth noticed, the parent bird stood up over the eggs, unsettled by the strange noise.

Front cover of the first edition.

Swifts in a Tower was published in 1956. Aimed at the general reader, it describes the birds' life at the nest and in the air. It had been out of print for years when I purchased a second-hand, second-edition copy with the wrong kind of swift on the front cover for £80, an alarming price for a small hardback. Still, the contents are pure gold. This small volume has remained the key reference work on swifts since its publication and has now been reprinted.

David and Elizabeth were married in 1949.

9

RINGS AND DEVILS

Oxford University Museum tower has proved an enduring observation post for swift behaviour, and two ventures have been of particular value: the British Trust for Ornithology's ringing scheme and Derek Bromhall's film, *Devil Birds*.

Over the last century almost 200,000 swifts have been caught in the UK, measured and given an aluminium identification ring, each with its own unique number. Nearly 5,000 of these are from Oxford, handled by Roy Overall in his forty-nine years as the tower's ringer. He followed David Lack's precise and careful rules: never ring an adult while it is nest-building, egg-laying, incubating or feeding its chicks as it may desert; ring them only when the young have fledged or a single adult enters the box. A fledgling should be ringed at five weeks old; leave it another week and you risk the bird erupting out of the nest prematurely, before its flight feathers have fully grown. This means that far fewer adult birds are ringed than young ones. Recoveries of ringed birds have revealed clues to the swift's migration routes and wintering grounds but, most powerfully, they provide evidence of the longevity of swifts. In 1964 Roy found a dead bird in the tower that had first been ringed as an adult in 1948, making it at least eighteen years old. Others have not been so fortunate. The remains – rings and legs – of three fledglings were recovered from a hobby's nest a few miles away just a few weeks after being ringed. There is great variability in the longevity of a swift; average life expectancy is thought to be around seven years.

Birds fledged from the colony seldom return as adults – less than 1 per cent have been recovered, though one ringed fledgling returned at least seven times. Those with an established nest, however, are absolutely site-faithful. Ringing has contributed tantalising hints about migration, but recoveries beyond Europe are vanishingly few, with less than twenty UK ringed birds recorded from Africa.

Today, information yielded by new tracking technology is calling into question the continued practice of ringing swifts. Geolocators fixed to a score or so of birds have revealed more about their migration routes and wintering areas than a century of bird ringing.

Roy took over a thousand journeys up the steep stone steps and the wobbly ladder and yet, with a merry twinkle in his eye, he claims not to have got to know swifts that well during nearly a half-century of monitoring the nests. 'It was always dark!' he says. This was a good thing in many ways as it hid the spiders of which he was fearful.

But there were other scary creatures here, too. The swift lousefly, *Crataerina pallida*, sends shivers down the spine of even the most hardened scientists. Lack described it as 'repulsive' and 'revolting'.

Lousefly on a recently-hatched nestling.

When at rest, this hairy, greenish-black flat fly holds its legs away from its body, giving it a spider-like appearance. It is entirely attuned to the breeding cycle of swifts. Unusually among insects, the single larva develops inside the body of the female. She deposits the pupa next to the nest, or in a crevice close by, where it lies dormant through the autumn, winter and spring, 'like a vitamin capsule in appearance', as Lack succinctly put it. The flat fly emerges as an adult when the temperature warms up in late spring. Sometimes this is when the swift nestlings hatch, other times at the moment when the parent birds return. Straightaway, it will fasten itself to a nestling (or an adult if the eggs have not yet hatched), from whom it will suck 25 milligrams of blood every five days. Three or four flies may crawl on to the young birds at once, occasionally up to a dozen. But except in the most severe infestations, the lousefly seems to have little effect on the health of the nestlings.

Its wings are vestigial; it simply has no need for them, spending the whole of its adult life attached to a swift, where it will feed and mate. One-third of an inch long, it is a formidable parasite; scaled up to human dimensions, imagine a four-inch monster fastening itself to your skin. It can move forwards, backwards and sideways. Try and squash one between your fingers and it simply slides away.

Crataerina's flattened body enables it to move freely among the swift's feathers and latch itself tightly to its victim. Its sharp, tridentate claws pierce the feather vane, and the barbs of that feather then become caught between the individual spines of its claws, exerting powerful force. Adhesive properties on the lobes between these fearsome claws, the *pulvilli*, make the insect difficult to dislodge. Roy remembers with horror how he had once attended an event at one of the university colleges: 'I was dressed in a suit and tie, very smart, then my wife Mary alerted me to a flat fly on my shirt collar. I flicked it off and stamped on it!'

He cleaned out the boxes before the swifts returned each year, but the nestlings invariably became infested. It may be that the pupae fall into cracks in the woodwork or that swifts carry the full-grown flat flies in from other nests. The fly's lifecycle is perfectly attuned to the swift's breeding period. Scientists have racked their brains for reasons why swifts might tolerate them. Could it be possible that they get something in return for all that blood-sucking? Nothing has been proved and the subject remains a mystery. The fact that the relationship is not detrimental to breeding success is strange; parasites generally do cause problems. For example, the nestlings of Alpine swift are known to suffer negative effects from *Crataerina melba* lousefly.

Roy also played a part in *Devil Birds*, working with Derek Bromhall on a film revealing the life of the swift as never before, its spell-binding story shared with a large new audience – including me. Broadcast in 1980 in the renowned *Survival* series, the museum's swift colony shot to stardom in the thirty-minute film and accompanying book.

Devil Birds is a wonderful mix of science and romance, capturing the drama and spirit of the birds' aerial lives, while also focusing closely on activity inside the nest. Narrated by Eric Thompson, whose voice conjures up memories of *The Magic Roundabout* and a slower-paced world, the film shows the swift as a quintessential part of an English summer, with sweeping shots over the crockets and spires of Oxford's roofs and the surrounding countryside – Blenheim Palace, cricket pitches and the local sewage works – then one of their best feeding grounds.

The book, a mixture of photographs and words, tells the story: 'In England the swift used to be called the Develing, Devil's screech, skir devil or Devil's bird; it is easy to imagine a pack of swifts, small black projectiles, hurtling from the sky at phenomenal speed and screaming like banshees, as a fistful of little demons flung by the Devil, returning to some Satanic roosting place.'

The front cover image alludes to this demonic spirit, with the head of a young swift against a dark background, its indigo eye slightly cloudy, shadowed by a deep brow. Handsome as Lucifer.

In the last days before it leaves the nest the young swift is at its most beautiful. Its long, pointed wings are fully grown, the primary feathers edged with white, deep-set eyes shielded against the friction of the wind under heavy brows.

Their devilish eyebrows have a highly-evolved purpose. I meet Derek at his home on the outskirts of Oxford. Amid a mini-rainforest of plants in his conservatory, with Chinese painted quail foraging close to our feet and diamond doves crooning in the branches – they sound a little like a child learning to play the recorder – he tells me how he came to make the film. A man of many talents, he had a varied career. As a student he won a scholarship to Sierra Leone, were he collected marine fish for the British Museum, each identified with its Latin, Creole or other local name. After graduating, he was appointed a biology lecturer at the University of Hong Kong, later joining the Hong Kong government as Director of Fisheries Research. In 1966 he returned to the UK and Balliol College, Oxford, to work on mammalian nuclear transplantation, the technique which led to the creation of Dolly the sheep.

He then made a career change and took up film-making. He wanted a challenge, something no one had tackled before. Oxford's Museum of Natural History, with its colony of swifts, presented him with the perfect opportunity.

Devil Birds was made with hard work and very basic equipment during 1976, the hottest summer since records began. 'It was like an oven in the tower, I spent most of the time working in my swimming trunks,' Derek

View from the tower.

remembers. For the nestlings, though, it went beyond discomfort. A number of those on the south side tumbled to their deaths as they desperately sought the fresher air at the nest entrance.

For the parent birds and for those nestlings on the marginally cooler sides of the tower, the ceaseless sunny days brought good fortune. Food was in abundance. The adults delivered so many meals they sometimes had to prod their chicks into eating them, so sated were they with hoverflies, plant bugs, aphids and spiders. A pair of adults may catch 20,000 insects in a day to feed their young; in 1976 this was easy.

Ingenuity was needed to film the swifts. In the nest boxes it was relatively simple, once the birds had become accustomed to bright lights. But certain shots proved difficult, even hazardous. Derek wanted to get footage of the birds entering the ventilation flutes and came up with a cunning plan. This involved rigging up a wire from the tower to the building opposite, from which a camera would be suspended. He tells the story vividly: 'To get the wire over, I had to fire it across with a bow and arrow. Julia, my wife, was standing below and shouted up when a swift was coming, so I knew when to activate the camera!' But the swifts were too quick for them, gone before the shutter could be pressed. Then Derek took to the air, hoping to film from a helicopter. A catastrophe was only narrowly averted when his heavy camera, suspended on a cable below the helicopter, was swept on take-off to within inches of the tail rotor by the draught from the helicopter blades. Somehow though, after two summers' filming, he finally got the footage he needed.

So it was that the life of the swift in its cramped hole was revealed to people on TV screens in living rooms throughout the country. The film, which won an award at the World Wildlife Film Festival in 1982, starts with the joyous ringing of their calls as they flicker and glide through blue skies, the city seen from their viewpoint and memorably described as 'a petrified forest inhabited by strange creatures'. The camera takes us into the narrow confines of their nesting places and we see the jerky, awkward movements of the birds as they shift about, their long wings an encumbrance on the floor – a strange contrast to their effortless aerial agility.

We watch a pair preening each other, see the moment of mating and witness the laying of an egg. A monster lousefly crawls into view, sucking blood from an adult bird; another latches on to a just-hatched, naked nestling. These young swifts are, as the narrator tells us, all beak and stomach, their enormous red gape demanding to be fed. Like spiky stubble, quills start to protrude through their skin. Then soft down appears among growing feathers. Finally, the ugly fledgling (as the script harshly refers to it) becomes a beautiful bird and slips from darkness into sunlight, leaving behind its shuffling existence in a tiny hole for boundless freedom on the wing.

In a prescient observation, tucked away in the acknowledgements section of *Devil Birds*, Derek Bromhall sounded a warning about the future of the common swift: '. . . it might be supposed that its life is so remote from ours that our influence on it must be minimal. But almost unawares we have been providing swifts with nest sites in the roofs of our homes for centuries; equally unwittingly we may be denying future generations of swifts their traditional sites, as new buildings replace old and roof spaces are sealed off to conserve heat. The swift is in fact in greater danger now than at any time in the past.'

He called for people to install nest boxes: 'And for those who take the trouble and have the patience to establish their own colony, there will be the immeasurable reward of being able to observe, enjoy and discover for themselves what it is about the swift that makes it the most fascinating and challenging of birds.'

That reward depends on the swifts choosing to make use of the boxes we put up for them. Derek installed some on his own house but they never moved in. For a few years starlings occupied them, fine chattering neighbours. Every summer he watches swifts feeding high above his garden but never again has he experienced the intimacy of those two summers filming swifts in the tower.

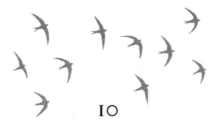

10

INTO THE TOWER

I am curious to find out how the Oxford colony is faring and what the research study, now in its seventieth year, has discovered. Most of all, I want to go inside the tower and see the swifts. In the early 1980s, during my time as a student, I had relished visits to the Pitt Rivers Museum, but for three summers had cycled past the building in blinkered ignorance of the birds' presence, while sharing the same air, rain and sun. Now, I am finally going to catch up with them. David Lack's son, Andrew, a lecturer in environmental biology at Oxford Brookes, has kindly arranged for me to visit. He will bring his wife, Helen, their thirteen-year-old daughter, Margaret, and bird ringer George Candelin. 'Wear dark clothes and sturdy footwear,' he instructs. Disturbance to the swifts is kept to an absolute minimum. So I pack a black T-shirt and jeans, along with binoculars and camera.

I take the train to Oxford. At Leamington Spa we transfer to a coach as engineering work is being carried out on the track. On arrival, I go to pick up my case from the hold. As one of the last passengers to join the coach, I expect mine to be one of the first out. But there is no sign of the small silver suitcase I had left there. Finally, the driver produces what appears to be my case, though a slightly duller shade than I remember. A second's doubt flashes through my mind, but I assume it is mine – there is nothing else left.

Needless to say, it is not. On lifting the lid, I discover layers of very neatly packed men's clothing, a laptop and a dissertation on the subject of bioinformatics, printed and bound for examination purposes. Eek! Instead of exploring the eccentrically informal gardens at Corpus Christi College as planned, I trail back to the station and alert everyone I can to the error. Phone calls are made to Banbury, where the coach has stopped, notes are written, phone numbers taken. My mobile stays ominously silent all evening.

I do not think the brown twill trousers in the case are really my thing – in any sense – so I arrive at the museum the next morning in the clothes I travelled in: a red T-shirt and an off-white skirt, along with black sandals. At least the sandals are the required colour, if not exactly sturdy. George is outside the museum when I arrive and, thankfully, he does not seem too worried by my inappropriate clothing. Then, at last, my phone rings: the man with my suitcase. Phew.

These days the museum takes pride in its swifts, and a large TV screen is showing footage of a rumpled-looking chick on a live webcam. It has not always been so. A former director regarded the swifts with deep suspicion, fearing the birds would bring an infestation of museum beetle, whose hairy larvae could devour the skin of the stuffed birds below.

There is quite a list of pests that museum keepers fear. These include the brown marmorated stink bug, cigarette beetle, firebrat, minute brown scavenger beetle, old house borer, vodka beetle and, of course, museum beetle. Natural history museums have, however, struck up a working relationship with one kind of insect. The larvae of *Dermestes maculatus*, a relative of the carpet beetle, are superb at cleaning the skeletons of birds and mammals destined to become exhibits in museums. Quick and thorough, the larvae consume the remnant tissues without damaging the bones. It is, of course, essential that they do not escape. If they were to get into the display cases of stuffed animals, they would soon reduce the exhibits to dust.

The chances of the swifts bringing any of these bugs into the museum are probably nil. Swifts have their own highly specific pests, but these have no interest in abandoning their hosts. The beetles would most likely enter on people's clothing or shoes.

George leads us to the foot of a gleaming metal spiral staircase. 'Actually, it's helical, not spiral!' says Andrew, with scrupulous precision. This sturdy structure has replaced the wobbly ladder. Two by two, in accordance with the health and safety notices, we skip up. 'So much easier than it used to be!' exclaims Andrew, stirred by memories of accompanying his father up the tower as a boy. Lack Senior shared his passion for birds with his sons, teaching them how to catch them in their garden using cage-style Potter traps. They would look closely at each bird before setting them free again. 'Red-backed shrikes and nightingales nested close by; it was an idyllic childhood.'

We arrive at the first floor with the glow of infra-red wall lights illuminating our surroundings, a not unwelcome modern concession to safety requirements. Slowly, our vision adjusts and George shows us some of the boxes, lifting the cloth from the Perspex window at the back so we can see the birds in their nests. From the first, a young swift peers out towards us, the white scalloping on its feathers clearly visible, that nightjar look about its eyes. Two pearly-blue pieces of confetti have been caught and woven into the nest cup. Nests vary but are generally the size of a small saucer. In one of these there are three nestlings, nearly full-grown; two at the bottom, the other on top of its siblings. This stacking of chicks is not unusual. In Lapland, where swifts nest in trees, people encourage them into their gardens with hanging, hollowed-out logs. Within, the chicks have often been observed one on top of another.

From an empty nest George picks up an eggshell. It looks strangely crumpled, its structure collapsed. 'Just before hatching, the chick absorbs calcium from the shell,' he explains. The question of how swifts get this vital mineral into their diet to build their bones and make eggshells is an enduring mystery.

Up a short, fixed ladder, we climb to the next floor and then once more to the top. Swifts seem to like the highest boxes best; these have the highest rate of occupancy. George tells us another curious thing: swifts appear to be right-handed. Handedness, or footedness, has been found in many animals and is well documented in parrots, in whom the left foot and eye are dominant. The nest boxes are nearly all semi-detached and, overwhelmingly, George has observed that the swifts choose the right-hand ones.

Up here, in the darkness of the tower, you find yourself in the swifts' world, distant from the rush of traffic and crowds of tourists. It provides a rare closeness to these birds, a perfect garret for academic study. We listen to the sound of the young birds, clamouring quietly for food, a soft, silvery version of the adult call. Each bolus of food is delivered to just one chick at this stage of their development, not shared as when they first hatched, so they need to make themselves heard.

Helen tells me that her mother-in-law, Elizabeth Lack, carried on watching swifts in the tower, scaling the tall, bouncy ladder until the week before her first baby, Andrew's elder brother, was born in 1952. Afterwards she gave up scientific research, devoting herself entirely to her family, as was expected of women in the 1950s. However, she returned to ornithological work when her children grew up, co-editing *A Dictionary of Birds* and taking a major role in the editing of the *Atlas of Wintering Birds in Britain and Ireland*. She maintained her interest in the museum's swifts throughout her long life, always eager to hear news of how the colony was faring.

From sixteen pairs at the start of the project, numbers grew to a high point of seventy-five in the 1980s and an annual average of 100 fledglings. This dropped gradually to eighty-one in the first decade of the twenty-first century, then plummeted disastrously in 2012. A particularly cold, wet summer proved dismal for insects and wretched for swifts trying to catch meals for their chicks. Nor were they the only hungry birds: a sparrowhawk took to hunting around the museum; perched on a pinnacle below the ventilation flutes, it would spring into the air as the birds left their nests, ultimately seizing all the adult swifts from the west side of the tower. That year the colony produced just fourteen chicks. Since then, numbers have gradually recovered. The final tally for 2016, the year of my visit, was twenty-six pairs raising thirty-six young. By 2019 the colony was looking considerably healthier, with eighty-five young fledged.

George worries, though, about the wider picture. 'I think that swifts are starving,' he says. 'Pesticides and intensive agriculture are wiping out their food.' Pinpointing the precise reason for the decline of swifts is surprisingly difficult. Here, at least, along rivers and over gardens, swifts seem to be finding enough food for their chicks. The global decline of insects is, however, evidenced in survey after survey and will inevitably affect them, especially where aquatic or other rich feeding areas are distant from the breeding colony. We can at least be assured that their nesting places in the tower are safe and hope that others will be inspired to make provision for swifts in their own houses.

II

MAY INTO JUNE

While breeding swifts are spending many hours in the darkness of their holes, spring has ebbed into summer and daylight stretches long into the evenings. The garden is swathed in green, my neighbours' houses hidden behind the vigorous foliage of a willow and a rambling rose so dense that a blackbird has built its nest among its thorny twigs. White and purple foxgloves are opening, bumble-bees crawling into their freckled flowers. The meadow has flashes of yellow, hay rattle flowers beloved by bees. Next will come buttercups and ox-eye daisies. The vegetation of early summer is lush and abundant, a source of food for myriad insects.

In the gentle hills nearby, along the Welsh border, not-so-common-as-they-should-be common blue butterflies are flickering over the grasses, the females seeking out the best plants of bird's-foot trefoil, drumming their feet on the leaves and listening, antennae bent down, to decide whether they are good enough to lay their eggs on; they want to be sure these plants will provide a plentiful supply of food for their caterpillars.

Another grudgingly named butterfly, the grizzled skipper, is resting on a wild strawberry plant. Its wings are patterned in black and white, a diamond of a creature. If you look closely at the right time you can find tiny, crescent-shaped bite-holes, where the caterpillars have munched at the edge of the leaves. I have a fondness for this plant too and gather half a dozen juicy berries, intensely flavoured and refreshing. Down among the trees, where honeysuckle hangs in twisted ropes, there is evidence of small-scale quarrying in the shape of an old hollow, with hundreds of hart's tongue ferns licking the cool air. As I stand and take it in, a sharp sound grates on my ear. I look round and catch sight of a blue tit, hopping from twig to twig and scolding me for my presence; it must have a nest close by.

I watch video footage, from a friend in Northern Ireland, of a swift entering a nest box on his house. This particular box was unoccupied last year, so the bird is likely to be a yearling. It looks around for a while, taking in the strange confines of its surroundings. Then it hobbles to the back of the box on feet that have not touched ground or hard surface since leaving the nest where it hatched. Tipping from side to side, its movements are awkward and its long wings, so graceful in flight, appear a hindrance. Like the train of a bride's wedding dress, they drag across the floor of the box; this bird is truly out of its element.

At last, though, it can preen itself. The swift buries its

beak in the feathers on its back and rubs at an itch that might have bothered it for months.

The breeders start adding a few feathers to the nests they made last year and sit companionably together. Their bond is renewed with hours of mutual preening; they are particularly attentive to those parts they cannot reach themselves, such as the throat. They make soft sounds to each other, *'les cris intimes, doux et discret'** as Swiss ornithologists have described it.

Breeding gets underway soon after the pair is reunited. Their sexual habits are legendary: the swift is the only bird known to mate on the wing. People were sceptical that this could occur, as Gilbert White noted in his account of aerial copulation in 1774. He urged those incredulous of its happening to use their eyes and witness it for themselves: 'If any person would watch these birds on a fine morning in May, as they are sailing around from a great height from the ground, he would see every now and then one drop on the back of another, and both of them sink together for many fathoms, with a loud, piercing shriek.'

For a bird so exquisitely adapted to the air, such breathtaking acrobatics are quite simply an expression of its way of life. Where their nesting place is particularly cramped, mating on the wing may be easier, but if they have more space or if it is raining, the nest may be preferable. During this activity is the only time when it is possible to discern for certain the identity of male and female birds. As he

---

* 'intimate cries, soft and hushed'

climbs onto her, he grips her back with his claws and the nape of her neck with his beak. She raises her tail, gives a soft scream and then they separate. This is usually repeated three or four times and is followed by another bout of mutual preening. Mating carries on throughout the nesting period, keeping the bond strong between the pair.

Common swifts are highly faithful to each other, though this is not always the case in high-density breeding colonies of birds, such as the fickle pallid swift. Research in Oxford's museum tower over a two-year period revealed a low rate of extra-pair fertilisation, suggesting that the stability of the pair bond is of greater value to the female swift than any possible benefits to her offspring resulting from extra-pair mating.

They can, however, be pretty quick at finding a new mate if one goes missing. This tends to result in fights if the original partner subsequently turns up.

It takes a great deal of energy to produce an egg. The finished product weighs 3.5 grams – one twelfth of the weight of the adult swift. So if the weather is cold, wet or windy and food is scarce, egg-laying is delayed until the weather warms up and airborne insects become more plentiful. The egg is almost always laid early in the morning. There is logic to this: carrying extra weight around when hunting for food is hard work; once the egg is out, she can head off and feed, unencumbered by the additional weight. Swifts lay two to three eggs, two to three days apart; the earlier they lay, the larger the clutch size. Like other hole-nesting birds, their eggs are white, making them easier to see in the dark. The parent birds share the brooding of the eggs equally over the incubation period – about nineteen days.

One of the most puzzling questions for ornithologists is the matter of how swifts get calcium into their diet, needed for eggshell formation. During the breeding period, female birds such as blue tits search for foods that contain this necessary mineral, such as fragments of snail shells. Sometimes you see them clinging to walls, pecking at the brickwork. This usually happens in the evening as eggshell formation takes place mainly at night. Swallows are known to supplement their insectivorous diet prior to egg-laying with a little calcareous grit. Dependent entirely on airborne food, this is not an option for swifts.

It may be that the aphids which form a large part of their diet contain traces of calcium derived from the sap of trees, drawn up from the ground through their roots. While the calcium is never metabolised into the aphid's own structure, it may still be present in its digestive tract at the point when a swift snaps it up. Perhaps they also ingest some from water sipped in flight from lakes fed by lime-rich waters.

Swifts also lay relatively few eggs, minimising the need for calcium. Single brooded with an average clutch size of two or three, their requirements are lower than those of a blue tit, which may lay fifteen eggs.

Things do not always go smoothly in the lives of breeding swifts, however. Sometimes they throw their eggs out of the nest and everything has to start all over again. This seems to happen when a male suspects the eggs are not his, perhaps because he has arrived back later than his mate. There is one much-watched swift box in Bristol where this happens each year. This bird clearly has an overwhelming suspicion of his partner.

Sometimes it is the female who picks them up in her bill and throws them out. This happens if her mate disappears and she is left alone. Rearing young as a pair is likely to be more successful, so her strategy will pay off. Soon, she will have a new mate and lay eggs again. Occasionally, the eggs are accidentally knocked out as the swift turns around in its nest. It has been known for the female to lay six or more eggs before finally managing to keep one or two in the nest.

Swift with eggs.

As soon as the first egg is laid, swifts will brood it at night, but until the clutch is complete the eggs are not incubated by day. In consequence, the first egg laid will hatch a little earlier, the nestlings emerging within a couple of days of each other. The pair take turns to do this. In bad weather, if the hunting bird is away for a long time, the sitter may get hungry and will sometimes leave for several hours to find food. Most small birds make only brief forays to feed themselves; the developing embryo is vulnerable to chilling, so if the eggs are left uncovered for long they will fail. Swift eggs, however, are resistant to cooling, except at the start of incubation, an adaptation that enhances the bird's chances of successful breeding in the variable weather conditions of our northern climate. Seabirds such as the fork-tailed storm petrel, which breeds in Alaska, have this adaptation too. Like swifts, they may have to fly long distances to feed.

When a chick is about to chip its way out of the egg, it first pecks through a membrane into the pocket of air at the blunt end. Here, its life reaches a new stage as it takes a lungful of oxygen; it is at this point that a soft piping sound can be detected before it emerges from the shell. David Lack heard it during his long hours of observation in the Oxford museum tower, but this sound will rarely be caught by the human ear.

A hatchling emerges from the egg with a foetal appearance: naked, with an enormous mouth and a huge gut, but not in the least helpless. Vigorously, it demands food, stretching its neck and raising its head towards the parent's beak. Its gape is huge, big enough for the parent bird to put its head inside and eject part of the food ball it has brought for its young. It will take five to eight weeks for the young birds to mature, depending on food availability. Like other hole-nesting birds, they are relatively safe from predators, making the nest a good place to grow. A black-bird, by contrast, is out of the nest in a fortnight – its nest is more exposed to raiders – but its parents will continue to feed it close by for a further three weeks. A swift is independent from the moment it leaves the nest.

Until the nestlings have feathers, their parents keep them warm by constant brooding, once again taking turns to go out for food. Unlike many birds, which bring in frequent, small meals, swifts collect 300–1,000 insects and spiders at a time, gluing them together into an enormous feast-ball, delivered every hour or two to their young. As it arrives with a meal, the parent's throat pouch is distended, as if it had half-swallowed a ping pong ball. Thrips, flies, aphids, beetles, spiders, spittle bugs, dung flies, moths, mayflies and sometimes small dragonflies are snapped up; hundreds of different insect species have been identified within the bolus.

When hunting, they fly relatively slowly – in swift terms. Cruising through the skies at around twenty-five miles per hour, they adapt their technique to the weather condi-tions, snapping up larger insects – 5–8 millimetres – when these are abundant, but hoovering up every tiny thing they come across when the weather is less favourable. For their chicks they will not bring anything too big, though they can catch and eat larger insects themselves.

According to David Lack, the swift can discriminate

between different prey, avoiding those that can sting, such as female worker bees and wasps, even when they are similar in appearance to harmless hoverflies and male drones. Some ornithologists doubt that this is possible, suggesting that swifts simply eject foul prey by shaking their heads with their beaks open.

The swift's eye is, however, well adapted to making shrewd choices. Like kingfishers, birds of prey and swallows, they have a double fovea. This is the bit near the centre of the retina, with the greatest density of light receptors. It is the area of greatest forward visual acuity, providing the sharpest, clearest detection of objects. The second fovea gives the eye enhanced sideways vision. Meanwhile, the eye is kept moist and clean of dust and debris through the action of a third concealed eyelid. This nictitating membrane (from the Latin word *nictare*, meaning to blink) is translucent, allowing the bird to maintain some vision as it sweeps across the eye like a windscreen wiper.

The timing of their hatching and forty-day growing period in the nest coincides with the longest hours of daylight, from the middle of June through July, with food gathering possible from around six in the morning until ten o'clock at night. The north has great advantages.

As the nestlings grow, each food ball is swallowed whole by a single chick. The young birds clamour for the returning parent's attention, calling and opening their beaks wide. In fine weather they put on weight fast, storing fat under the skin, which keeps them going in leaner times. Each day the parent birds will bring an average of forty meals for their chicks, a total catch of some 20,000 insects.

Just as the eggs remain viable through cold periods, swift nestlings have evolved a way of dealing with cold, wet weather. Unlike most birds, they can become torpid, conserving their energy until conditions improve and food becomes plentiful again. Their body temperature may fall dramatically at night, slowing down their functions and limiting energy expenditure.* By day they become active again – alert and ready to be fed, should their parents come back with food. Songbird nestlings will die if no food arrives within a few hours; once they are a few weeks old and have fat reserves, swifts can survive up to five days without nourishment.

Torpidity of nestlings also helps their parents. When local supplies of food are scant in poor weather, the adults can look after themselves, flying great distances to find better conditions and insects. In this way, the chances of both adult birds and chicks surviving are improved. When the temperature warms up and food is abundant again, the young chicks rapidly catch up on weight loss.

---

* The swift's relative, the hummingbird, also uses torpor to survive cold nights. Adults can suspend their usual metabolic rates, enabling these tiniest of birds to reduce heat loss overnight. Deep torpor sets in: 'No motion of the lungs could be perceived . . . the eyes were shut, and, when touched by the finger, [the bird] gave no signs of life or motion.' Alexander Wilson, 1832.

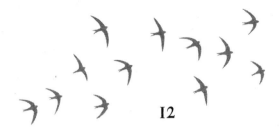

12

A SECOND CHANCE

May, June and July: in these months we revel in longer days, warmer temperatures and the leafiness of summer. Not, though, if you are a swift rehabilitator. Around the UK there are a number of people who give up their summers to look after the casualties of the breeding season, offering a second chance to swallows, martins and swifts. It is specialist work. I had phoned one of them, Gillian Westray, when I looked after my swift, to ask for advice. She'd sounded a little harassed: 'I can't talk now, I have sixty hungry birds to feed! Can you ring later?'

The mortality rate of many wild creatures in their first days and weeks is very high. Inexperienced in finding food and often without the wariness essential for survival, they easily fall to a range of predators. Cruel though this seems, it serves an ecological purpose, providing food for other wild creatures. More than 88 per cent of robins, for example, will die in their first year, some predated in the nest, many others within days of fledging. Just one in a thousand eggs from frogspawn may make it to adulthood. Fortunately, reproductive superabundance ensures that enough survive to keep the generations going, expanding when conditions are favourable and contracting when their environment deteriorates.

A brief life may be the certain and natural fate of many young creatures, but when people find a young bird that has perhaps fallen out of its nest or been brought in by a cat, their instinct often tells them to try and do something to save it. Most fledglings, however, are best left alone; many birds carry on feeding their young for days or even weeks after they have left their nest and will return with food. For swifts, though, this is not the case. When a swift falls out of a nest or fails to get airborne on its first flight, no parent bird will come to feed it, so a grounded fledgling really is in trouble.

Gillian has raised over 1,000 swifts over the years but is still learning about them:

Every bird is different, each has its own personality. This is true of all birds, but particularly so for swifts. I have tended small, lively vocal ones and big, laid-back sleepy ones; surly ones which keep to themselves – these are always the quickest to be off; others that love to be handled and seem to find it hard to leave. Yet despite the intimate relationship between carer and bird, swifts retain their wild instincts untamed; the migratory urge is never subdued, whereas swallows and house martins will imprint on the people who look after them, more or less turning into pets if you are not careful.

Her foundling birds are given meticulous care. At the heart of this is diet. Scientific evidence shows the necessity of insectivorous food for hand-reared birds. Research carried out in Spain by Enric Fusté demonstrated that birds put on less weight and their plumage is poor when given a non-insectivorous diet. The birds were fed either rat mince, using laboratory-reared, pathogen-free animals; kibble, a formula based on a high-protein–low-carbohydrate cat food; house crickets and wax moth larvae; or mealworms. Their progress was measured against weight gain recorded for wild fledglings.

Both insect diets performed comparably with the progress of wild birds, but the non-insect diets showed significantly poorer results and led to stunted growth. Insectivorous birds digest their food more quickly than carnivorous ones; it is an efficient system that maximises the nutrition in their natural diet. Meat requires a slower digestive process, so when common swifts are fed a carnivorous diet, their bodies have a briefer opportunity to assimilate and metabolise the food completely.

Swifts eat a vast range of insects, absorbing different nutrients as they do so. When hand-reared, the food they are given needs to mimic their natural diet as closely as possible or they may become deficient in minerals, which can lead to weakened bones and problems in the development of feathers. Common swifts need to be in exceptional body condition from the moment they fledge, so getting their diet right is crucial.

As Fusté writes: 'We cannot provide for the needs of orphaned common swifts as their parents do, but we must emulate them as closely as possible if we want to give them any chance at all of survival, initially for the long migration journey, and then to reproductive age.'

Raising a swift takes time, dedication and knowledge and is not for the faint-hearted – as I had discovered. At the end of each season, Gillian vows never to do it again – 'it's a form of insanity!' she says. But when the swifts return and the first casualties appear, she takes them in without a moment's hesitation. At the end of this summer, however, she is adamant that she will stop, at least for a year. She has relatives who need looking after and her house is falling down around her, she tells me. She sounds exhausted. Once again, this summer she has raised more than eighty swifts, swallows and martins brought to her by the RSPCA and other individuals, giving these birds a second chance.

It is not the birds that are wearing her down, however, but some of the people who contact her. She sometimes gets phone calls at three in the morning to ask for advice. It is as if she is expected to provide a 24-hour helpline, seven days a week. While most people are appreciative of her advice and understanding of the constraints on her time, she is also increasingly beset by rudeness in callers and an unwillingness to listen; people who will not read the guidance on the internet about what to do and, even more vitally, what not to do with injured swifts.

Judith Wakelam also devotes her summers to swifts, swallows and martins. Like Gillian, she lives on her own and is able to concentrate entirely on the birds brought to her. Neither of them takes holidays during the summer, not even days off. Each year, for four months, they devote themselves completely to the care of these birds, which would otherwise die.

Her love of animals is immediately apparent when a lively German shepherd dog bounds up to greet me at her home in a village in Cambridgeshire. Small birds jostle around the many feeders in her garden.

'I remember bringing home all kinds of ailing creatures as a child. My mother encouraged me – except when I came through the door with a nest of tiny rats: "take them back to where you found them!" she instructed.'

Little surprise, then, that many years later, when she spotted a tiny bundle of black feathers on the ground while out walking, she looked carefully. It was a young swift. Back home, she set about finding it food, catching moths and hoverflies from her garden. 'You won't rear a swift!' everyone told her. She rang the British Trust for Ornithology who put her through to Chris Mead, then one of the UK's top swift experts and an early champion of swift boxes. He had never succeeded in raising one, but encouraged her to try. 'Waxworms! Buy it some waxworms!' he suggested. For over two weeks Judith fed and nurtured it. And then it flew.

Since then, she has rehabilitated many more young swifts, brought to her from all over eastern England, forty in the last year alone. Judith confides that before she retired, 'if I had very young birds I would take them to work with me in a box and put them in the stationery cupboard so I could feed them often. Some of my colleagues wouldn't have approved, but they never found out!'

Once she took a call from some builders who had found a nest of young swifts while carrying out a roofing job. A twenty-four-hour experiment to see if the parent birds would find their nestlings in a hastily fixed swift box failed, so Jake Allsop, one of her Action for Swifts friends, brought them over to Judith in his hat. As they grew, she sent photos to the builders, so they could see the progress of the young birds. 'Strictly, they could have been reported for disturbing a nest site, but if I had done that, they would never have tried to help again. As it was, they were thrilled to have saved these birds.'

Releasing her swifts is always a wonderful moment. 'On one occasion I took six down to the local cricket ground just before a match was due to start – causing a bit of a delay. The visiting team from Croydon were fascinated, watching as the birds took to the air – the captain said he had never seen a swift before and the team gave them a round of applause.'

Few people get such a close-up experience, but these cricketers may perhaps look up into summer skies more often and watch the swifts' awesome aerial flypasts.

All Judith's young birds are ringed and she longs for the day when one of them is found breeding in the church tower, where a growing colony is present, thanks to boxes installed by friends. These were fitted when a cottage near to her home, where numerous swifts nested, was demolished and replaced with two new, perfectly sealed houses. Now there are twenty-six pairs in All Saints' Church, and forty swift chicks have recently fledged. One way and another, Judith and her swift friends are finding

Judith Wakelam releases a swift.

ways to ensure swifts keep coming back to the big skies of East Anglia.

With swift numbers in steep decline, the work of specialist swift carers in the country has real poignancy. The value of their work was thoughtfully described in an article by Helen Macdonald in the *New York Times*: 'Against a backdrop of environmental destruction and species decline, anxieties about our impact on the natural world become tied to the tragedies suffered by individual animals . . . Tending animals until they are fit to be returned to the wild feels like resistance, redress, even redemption.'

We are not alone in our efforts to rehabilitate wild birds; in Spain, for example, there is a wildlife rescue centre in each of its forty-two provinces. Not all of them accept swifts. Rural provinces can be overwhelmed by casualties from forest fires, and when this happens priority is given to the rarest species, such as eagles and falcons. Many swifts are brought to and tended in the urban centres. For example, in Barcelona, in the rehab centre where Fusté has worked for ten summers, more than 2,000 swifts may be raised and released in a single year. He is passionate about his work, but before I can raise the question, he says that, even on this scale, it is not conservation. His scientific training spurs him to question the validity of this activity: are these bird casualties the ones that nature would have let die? Most of the birds brought to the centre are under-nourished. Were they already weak in the nest, not begging for food? 'We are doing this for ethical reasons, treating wild birds as we would ourselves – as individuals; that is not how nature works.'

A delightful story comes from the renowned colony of Erich Kaiser in Germany. 'The Kaiser', as he is known to his international swift friends, started attracting swifts to his house in 1965. He now has around thirty breeding pairs and has observed the activity in his colony for more than fifty years. One pair of swifts would return each summer, refeather their nest, preen and go through all the rituals of mating, yet no eggs would be laid. Nonetheless, the pair stayed together. One year Kaiser discovered an abandoned nestling and gave it to the eggless birds to foster. For seventeen years these birds have returned to the colony and, each year, after it is clear that they will not be laying eggs, he has provided them with foster chicks, which they rear as their own. Other people are trying this approach too, adding young, abandoned birds to existing nests with one or two young. Initial results have been encouraging, with nestlings quickly accepted both by foster siblings and parents. This seems a sensible approach, the birds reared naturally.

Yet there are many questions about this intervention. Might such interference cause the parents to desert? Will they be able to gather enough food to feed their enlarged brood sufficiently? David Lack noted that some nestlings in the Oxford museum tower colony died most years through the failure of their parents to bring them enough to eat, and that among larger broods mortality rates were higher than those in small broods. In a decade of observations he recorded a 10 per cent lower success rate in nests of three chicks than those with two. In cool, wet summers birds may struggle to feed their own young and an extra nestling will mean less food for all of them. Only in the finest summers did broods of three all fledge successfully. Intervention in the life of a swift or any other breeding wild creature should never be undertaken lightly.

What I hope for is a time when the decline in swift and other bird populations is halted, when their numbers will rise and grow strong again, when we will run the world with the needs of wildlife in mind, with nature twined through our lives in towns, villages and cities; when its necessity, value and wonder will be recognized and integrated into the strategies of industry, agriculture, parks and gardens. When this comes about, as I dream that it might, it will no longer seem so urgent that we try to save every injured bird we come across. There will, I hope, always be people like Gillian and Judith who give up their summers to save individual birds, their work a manifestation of humanity at its best, but we will also be secure enough to accept that the death of young, wild things is part of nature.

What to do if you find a grounded swift

Looking after a swift or any bird is more demanding than you might think. First of all, it requires a big commitment of time, especially when taking on younger, less developed birds. Getting their nutrition right is essential. In addition to providing a diet of insects, rehabilitators give them vitamin and calcium drops that naturally-reared birds would get from the saliva that glues together the food balls brought by their parents. Without these supplements, their feathers will not develop properly.

Swift rehabilitators wring their hands when people bring in the often wretched birds they have tried to raise themselves. Sometimes the sheath of a feather may be an inch long but the feather itself barely developed; other times the feathers may be long but weak and brittle. Nor is it only have-a-go individuals that get things wrong; in some cases it is vets or wildlife rehabilitation centres. Curiously, there seems to be considerable resistance to accepting the science of swift nutrition. It is easier and cheaper to rear young birds on minced meat, but it is absolutely not in the birds' long-term interests. The signs of damage are not always immediately obvious; they may fly, but their plumage, which needs to be in perfect condition, is unlikely to be sufficiently strong to withstand the pressures of constant flight.

Experienced swift carers generally advise people against trying to look after the birds themselves. However, there are some basic rules to follow if you do find a grounded bird:

- Put the bird in a box with several air holes, on kitchen paper, in a quiet place for a few hours or overnight.
- See if it wants a drink by soaking a cotton wool bud in water and wiping it along the side of its beak, avoiding the nostrils.
- Don't feed it.
- Don't throw it in the air.

If it is early in the season (late April–May) the bird will probably be an adult. After a rest it may well recover and fly. Take it to an open space (providing the weather is favourable), preferably on a hill, and hold it flat on your hand above your head. If it is ready to fly, it will. If it falls to the ground, don't try a second time, but seek advice from a specialist swift carer.

If it appears to be a young bird, place the chick in a box (as above) and contact a specialist swift carer as soon as possible. If you can, send a photo and the bird's weight. Do not try to feed it, as the wrong diet can cause irreversible damage to feather growth.

13

SWIFTS IN AFRICA

The life of the breeding swift has been intensely scrutinised. In recent decades the availability of webcams has allowed many people outside the academic world to observe fights, mating behaviour, egg-hatching, chick development and monotonous hours of incubation, when the odd twitch of an itch or some rapid preening is the only action. The majority of its life, however, is spent in sub-Saharan Africa, away from its breeding grounds. About this period we know very little.

For centuries, common swifts have been observed in many parts of Africa during our northern winter, yet until recently the details of their journeys have eluded us. Ringing schemes have resulted in hundreds of thousands of swifts being issued with metal identification bands in their breeding grounds over the last hundred years, but the number of recoveries from Africa is minuscule. The chances of someone finding one of these tiny rings, appreciating its significance and then posting it back to the

British Museum (a website address is now also embossed on the metal) are remote indeed.

Of nearly 200,000 swifts ringed in Britain and Ireland between 1911 and 2015, just thirty-four have been recovered on the African continent. Few though they are, these records offered tantalising clues as to migration routes and destinations. Eighteen were recovered from the Congo Basin, eleven from Malawi, two from Tanzania and one each from Zambia, Zimbabwe and Mozambique. From these scattered records and from direct observation on the ground, it was deduced that swifts spend months across a broad sweep of Africa, from the skies above the Congo River and rainforest, through eastern Africa and onwards to the mangrove swamps of the Indian Ocean.

Constantly on the wing, they range over many countries, taking advantage of the rainy seasons that shift north–south and back again during the year and the bountiful insects that erupt after downpours. Swifts follow in the wake of the great wildebeest migration that loops through the vast plains of the Serengeti in Tanzania and Maasai Mara in Kenya, flying low to snap up the flies that thrive on the beasts' dung. They skirt thunderstorms, feasting on swarming termites, protein-rich insects that evolved from cockroaches and which nourish hundreds of different creatures, from aardvarks to eagles. Over the grasslands of the Sahel they forage high, guzzling insects whipped up into the air. Above rainforests there is good foraging and an array of insects, which multiply rapidly in the tropical conditions.

While adult birds may start to moult their body plumage when still in their breeding territories in late summer, most of the process takes place during the months in Africa. Keeping their plumage in top condition is essential for flight, and adult swifts, like many other birds, replace all their feathers each year. Feathers are strong and durable, and the melanin pigments in swifts' dark plumage provide better protection from the sun's rays than birds with pale feathering. However, constant exposure to wind and weather is damaging; sunlight breaks down the proteins, causing the feathers to grow brittle. Growing new feathers is energy-consuming, but food is abundant in Africa. First-year swifts do not replace their wing feathers until their second autumn, but new body feathers will grow, pushing out the weaker ones, which are characteristic of the plumage of young birds.

Common swifts are less vocal in Africa than in their breeding grounds, often foraging silently in loose groups, sometimes hundreds together. Occasionally, though, they are heard. James Wolstonecroft, a bird tour leader with long experience in Africa, tells me how he saw a flock of swifts drifting around offshore before rising up for the evening roost, calling softly to each other. Sometimes, too, he hears them as they feed.

Flying in groups gives them some protection against predators; if a raptor is spotted, a warning cry from one bird will alert the whole flock. Large gatherings of swifts also provide a signal to others of good feeding areas, perhaps a swarm of flying ants or insects disturbed by bush fires.

Swifts have adapted to cope with a global climate range, regulating their body temperatures in different ways. Just as chicks become torpid to survive cold weather and poor food-gathering days, these birds have evolved ways of coping with extreme heat, such as they encounter in Africa. Their legs and feet are normally folded tight to the body, to keep them streamlined and energy-efficient, but in temperatures of 30 °C or more they trail their legs and spread their toes to expose them to the wind. The toes and backs of their legs are not feathered and the front has just a slight covering, enabling them to lose a little body heat. Occasionally, they will also gape (as chicks sometimes do during hot weather). However, this strategy carries the risk of unselected aerial matter getting into their mouths, while in arid conditions it could lead to dehydration. This behaviour was first described by an ornithologist during a hot summer in Berlin.

Our ability to discover the details of birds' migration journeys has been transformed by the invention of the geolocator. The first generation of these light-logging devices weighed more than 100 grams and were suitable only for tracking large animals. Technological advances have now led to tiny devices weighing less than a single gram, making possible the tracking of even small birds. While sparse ringing recoveries provided some indication of the birds' travels, geolocators are extending our knowledge by leaps and bounds and bringing much greater precision to the map of swift migration. Passage routes and wintering areas are being identified and migration speed revealed.

Swedish biologists at Lund University have led the way in swift migration research, with a trial first taking place in 2009. Subsequent studies have added rich detail to the story of their journeys. They initially tracked six individuals through a complete migration cycle from Sweden to Africa and back, with data logged twice a day. During the autumn journey, all the birds took a southward course through Europe – with some variation of the precise route – followed by a shift south-west through the Western Sahara to sub-Saharan stopovers, then a swing south-eastward to their final wintering areas in the Congo Basin. This region in West Equatorial Africa is one of the most biodiverse areas of wilderness left on Earth, with forests, rivers, swamps and savannah all providing rich feeding opportunities. The swifts' journey here from Sweden took an average of sixty-nine days, including thirty-nine days of stopover feeding.

After approximately six months roaming over rainforest, spring migration began in late April. This journey was completed much more quickly than the autumn migration, but for five of the birds it involved a highly significant detour and stopover. While one bird flew more or less due north, the others swung north-west, flying over the forests of Liberia for an average of seven days. This was news to ornithologists, for no ringed swift had ever been recovered from this country and it was believed that swifts travelled without extended stopovers during the spring migration. A stopover for a swift is, of course, a slight misnomer. They do not cease flying but range continuously over an area, flying hundreds of miles. There are, however, good reasons why they would choose this longer route. Their arrival over Liberia coincided with the rainy season and the mass emergence of flying insects, including flying ants and termites, a feast that would not only keep them going on their migration journey but also sustain them for the first week or so when they arrived at their destinations, should it be cold and wet – as is quite often the case in early May.

Considering the whole life of a species is essential for effective conservation. Knowledge of migration routes can, potentially, inform strategies to help protect the birds' stopovers and wintering areas. Inevitably, this is difficult. According to Global Witness, a charity that campaigns against environmental abuse, Liberia is home to 43 per cent of the remaining Upper Guinean forest, which also covers parts of Guinea, Sierra Leone and Ivory Coast. For many years its forests have been logged to fuel the wars waged by its leaders. More recently, a new government has instigated a series of reforms to help this poorest of countries back on its feet. Sustainable community forestry is at the heart of these measures, and international financial backing, in particular from Norway, has encouraged the adoption of this approach.

Unfortunately, while reform programmes have been initiated, they have so far proved ineffective within the forestry sector. According to Global Witness, all of Liberia's logging contracts are illegal. Timber continues to be extracted by huge international companies, potentially causing irreparable damage to Liberia's rainforests. Responsibility for this failure to implement the reforms lies not only with the country itself but also with those importing the timber. For example, while the EU has strong regulations in place prohibiting companies from buying illegally-produced timber, the rules are not always enforced. Logging companies are adept at reinventing themselves with new identities, all too often evading scrutiny.

The destruction of Liberia's forests is disastrous for the people who live in them – about a third of the country's population – and devastating for its extraordinarily rich biodiversity, which includes 695 species of birds, both resident and migratory. Some stay for just a few days, feeding up on the journey to and from their breeding grounds, such as our swifts.

Perhaps for thousands of years, swifts have taken this route on their twice-yearly migrations. It is less direct but offers great rewards. The geolocator on the bird that skipped the detour and migrated directly north showed the slowest migration speed (which excludes stopover time), despite a shorter journey. Headwinds as it crossed the Sahara demanded more effort than the westward route taken by the other birds, which enjoyed favourable winds, especially at high altitudes. Wind patterns in the Sahara during spring are generally stable, a fact that confirms the ecological advantages of the five swifts' westerly route. Ornithologists were thrilled to have learned this new detail in the story of swift migration. Aware of the high rates of deforestation in West Africa, however, some of them are also worried.

The spring journey is a much less leisurely affair than

the autumn one, with far fewer days spent on stopovers. On average, the six swifts involved in the trial spent just eight days refuelling, compared with thirty-nine on the autumn migration. The birds have a strong incentive to get back to their breeding areas quickly. A late arrival puts them at risk of losing their nesting site or even their mate. Swifts make the journey rapidly for their size, thanks to their streamlined body and long, narrow wings, which maximise energy efficiency. Their ability to forage while flying also assists them on their journeys.

A year after the Swedish project, British scientists conducted a similar trial, with nine swifts fitted with tracking devices in the UK. They followed a similar route to the Swedish birds on passage but travelled even more extensively through Africa. The journey of one individual, identified as A320, shows it leaving Norfolk on 23 July, sweeping south-west through France and Spain, continuing over the forests of West Africa, then swinging south-east to the Congo Basin, where it remained for four months ranging over steamy forests, glades and grassy river banks, before heading south-east again through Malawi as far as Mozambique. Here it roamed for four weeks, before starting its return journey on 24 January. A320 then appeared to spend another two months over the Congo Basin before departing for its breeding grounds on 5 April. It followed the same route back as the Swedish birds, sweeping west and looping over Liberian forests for ten days before heading north. A320 took ten days to complete the last 3,000 miles of the journey, but another of the tagged birds took just five days – it probably had stronger tailwinds. These birds had flown more than 14,000 miles since leaving eastern England the previous summer. Our birds are truly nomadic.

The annual common swift migration route.

Common swifts converge on sub-Saharan Africa from across their far-flung breeding grounds, their arrival and departure determined by the timing of seasons in their nesting areas. Epic journeys are undertaken by all these birds but none more so than by *Apus apus pekinensis*, a sub-species found in Central Asia and China. It is slightly paler than *Apus apus*, with a softer call. The first known visual representation of this swift was found in artefacts in a royal tomb dating back more than 3,000 years. At that time, Chinese people believed their ancestors were transformed into swifts after death, so these birds have had a special place in their culture ever since.

As elsewhere around the world, local ornithologists had noticed that the number of swifts circling the skies over Beijing had fallen dramatically. Traditional buildings with cracks and holes have been replaced by modern, sealed developments, inhospitable to swifts. It is estimated that their numbers have dropped an alarming 60 per cent in the last thirty years. As with European swifts, their migration routes until now had been a subject of speculation, supported by fragmentary evidence. In 2014, scientists from China, Belgium, Sweden and the UK hatched a plan to fit microdata-gathering tags to swifts nesting in the Summer Palace heritage site, where they were known to have bred for several hundred years.

The Chinese ringers placed nets around the pagoda housing the study colony at two-thirty in the morning and, with plenty of skilled hands, thirty-one geolocators were attached to known breeding birds at first light. The ornithologists reconvened in Beijing the following year to retrieve the light-loggers. They succeeded in capturing thirteen of the birds involved in the experiment with the devices intact. The secrets of their journey were now at the ornithologists' fingertips.

Beijing's swifts begin their long journey to Africa in late July, taking a route west-north-west into Mongolia, then north of the Tian Shan Mountains, south-west through Iran and central Arabia into tropical Africa, arriving in Namibia and the Western Cape, where they spend three months before returning via a similar route in February, arriving in China in April. Flying constantly, there is neither beginning nor end to their journeys. This

particular one involves an aerial circuit of more than 16,000 miles each year.

The awesome story of their migration became a springboard for action in Beijing. Children were inspired to campaign for swifts, and many nest boxes were installed on school buildings. One young girl had bigger ambitions. She suggested contacting China's leading property developer, to persuade him that swift bricks should be installed in all his new buildings. A meeting was arranged between Mr Pan Shiyi and the schoolchildren. The young swift ambassadors gave a presentation, showing the swift's migration journey and its uniquely aerial life. Encouragingly, he gave them several swift boxes he had made himself, using offcuts from building sites.

Better still, he told them, 'Our focus has always been on humans and how to make our lives better. In the future, we need to consider biodiversity and to create a better living environment for both citizens and wildlife, such as the Beijing swift.' Will these fine words be followed up with action? The children can be trusted to hold Mr Pan Shiyi to account.

14

ACTION FOR SWIFTS

Scientific reports on wildlife decline in the UK and around the world make the headlines with grim regularity. I will be brief and selective – I do not wish to scare anyone away. In the last thirty years, the total number of birds singing and winging their way from tree to hedgerow, field to woodland, reed to rush or sky to roof is estimated to have dropped by 421 million across Europe. Losses are spread across many species. So-called common birds, such as sparrows and starlings, sustain the majority of the losses in both abundance and distribution. Rarer species tend to be held back from the brink of extinction by statutory conservation measures. Agricultural intensification, climate change and urbanisation are taking a massive toll on bird populations.

One of the most worrying signs of our degraded ecosystems is the calamitous reduction in insect life recorded in recent years. For example, total insect biomass in 63 nature protection areas in Germany fell by 76 per cent between 1989 and 2016, with a particularly alarming midsummer decline of 82 per cent. Nature reserves tend to be isolated fragments of habitat, too small for insect life to be unaffected by the toxic chemicals so liberally applied across intensively farmed fields. Entomologists in the UK are seeing the same patterns here, with the abundance of even supposedly common species shrinking every summer. How many hoverflies and soldier beetles have you counted on a head of umbellifer flowers in recent summers? It used to be dozens, now it is more likely to be three or four. Do you know of more than a handful of meadows where you can listen to the summer chirruping of grasshoppers? At the same time, three-quarters of the UK's butterflies are in long-term decline.

Agriculture's dependence on pesticides is disastrous for invertebrates. Nature reserves need to be enormous if their insect populations are to withstand the effects of chemicals applied on crops in surrounding fields. Of particular concern in recent years, neonicotinoids are now known to be causing serious harm to bees and butterflies. Numerous scientific trials, including those conducted by the pesticide industry, have shown how honey bee populations are reduced by exposure to these neurotoxic chemicals. Honey bees are renowned as one of the most robust of insects, so the effect on less resilient species is likely to be even more deadly. The regulatory system requires manufacturers to test on only seven prescribed invertebrate species, so there is little available information concerning their impact on hoverflies or other staple creatures in the swift's diet.

In addition to neonicotinoids, farmers may spray non-organic crops a dozen or more times while they are growing, with anything up to twenty-three different chemicals. The combined effects of these pesticides is even more toxic than each of them individually.

Peter Melchett, former Policy Director of the Soil Association, wrote: 'Neonicotinoids are supposedly highly targeted insecticides yet the researchers have found that they are turning up in the pollen of poppies, blackberries and hawthorn blossom in hedges, at levels that on their own are enough to cause harm to bees. Worse still, they are present along with a whole cocktail of chemicals, some of which could increase the toxicity of neonicotinoids up to a thousand times.'

According to the Department for Environment, Food and Rural Affairs (Defra), farmers in the UK sprayed, on average, 17.4 applications of pesticide to each hectare of arable land in 2015, using 16,900 tonnes of active ingredients. Imagine what this means for those creatures that depend on a good supply of insects, such as bats, hedgehogs and all our songbirds – even seed-eaters such as goldfinches need insects mixed in with the seeds they give to their nestlings. Crashing insect populations are wiping out the food supply of insectivorous birds, quietening the dawn chorus in hedgerows, woods and gardens. The threat of a silent spring is very real.

Writing in 1977 about technical efforts to improve the lot of man, such as the use of pesticides, Henry Douglas-Home commented poignantly: '. . . they have been made with little consideration for those fellow creatures who have already lived for so many millions of years before we developed our supremacy of mind. Before it is too late we must realise that we hold the keys of destruction – not only of ourselves but also of the more ancient occupiers of the world.'

We know well enough what we are doing, but fifty years on, the age of the Anthropocene has simply tightened its destructive grip on the natural world. Wild creatures have to contend with myriad pollutants, from the emissions of industry and traffic to plastic. While albatrosses choke on the plastic that floats across the world's oceans, people and every other living creature are also contending with the consequences of this durable material. Plastic fragments turn up everywhere, from the Arctic snow to the air we breathe, drifting on wind currents, settling on land and water. It has been shown that mosquito larvae consume these tiny particles and that they persist in their gut when they emerge as winged adults. What effect might a build-up of such plastics have on swifts and all the other creatures that eat mosquitoes?

Little wonder, then, that insectivorous birds in particular are struggling. However, it has been argued that swifts, being superlatively mobile, are better equipped to withstand the losses than other species, since they travel long distances, apparently effortlessly, to find food. Evidence is mounting each year, however, that they too are in trouble. Our damaged world is no longer an easy place for them to make a living.

Swifts are notoriously hard to count, so gathering reliable

statistics for the stability or otherwise of their population is difficult. Nest site recording gives accuracy but is tricky, as the nests are hidden from view and time-consuming to find. Counting birds in screaming parties, flying low over roofs, is a useful technique, but how do you know you are not recording the same birds twice or three times on different streets? It takes a highly experienced swift surveyor to be sure.

Some countries work harder at data collection than others, but evidence of a steep drop in numbers has been gathered by bird-watchers in the UK. In the twenty-two years between 1995 and 2017, swifts declined by 57 per cent. In the ten years between 2007 and 2017, their plight grew worse, with a 42 per cent fall in numbers: an annual average decline of 5.3 per cent. The latest figures released by the BTO show the number of breeding pairs in the UK has dropped to just 59,000: a 32 per cent fall in just seven years. The way we live our lives and shape our world is becoming less compatible by the day with the needs of swifts and many other wild creatures. In Scotland and Wales the twenty-two-year decline is even steeper.

Other European countries report losses of 25 per cent or more over the last twenty years. The European Bird Census Council (EBCC) has noted a moderate decline from 1980 to 2005. Oddly, their graph shows a slight recovery between 1996 and 2005 – the very time when surveys in the UK start to record a steep fall. Widespread anecdotal evidence suggests a downward trend across the continent, but while the serious decline in the UK has been recognised by the International Union for Conservation of Nature (IUCN) with an endangered designation for our swifts, the European population is still, officially at least, considered stable.

However, Birdlife International, a nature conservation partnership of 120 non-governmental organisations, has designated the swift a species of least concern, largely on account of its vast global distribution. Native to 124 countries, from Mauritania to Mongolia and Norway to South Africa, the world population is estimated loosely at between 95,000,000 and 64,999,999. Look at the swifts' fortunes in different countries, though, and the picture is not quite as rosy as these statistics suggest. In the UK in particular, there is growing evidence of a continuous decline. And across Europe there are concerns that common swift numbers are shrinking. Given that Europe is believed to hold 40 per cent of the global breeding population, there is strong reason to be disturbed.

Why has it taken so long to raise the alarm that our skies are growing quieter and these dark racing birds scarcer? Do we take action only when a species slips into the dreaded red category? That is indeed the moment when official protection measures are activated, when strategies for recovery are drawn up and legal requirements introduced. Such designations are not made lightly; a continuous downward trajectory has to be recorded for twenty-five years. So, while there was already strong anecdotal concern that swifts were in trouble in the 1990s, there was insufficient scientific data available to support their legal protection.

Why too did so few people, many of whom share their

houses with these birds each summer, fail to notice their absence? That is an easier question to answer. Like me just a few years ago, most people are not tuned into their screeches; their aerial acrobatics go largely unnoticed. There is so much noise from our own world, so many distractions to snare our attention and discourage us from looking up. Besides, there are still more than 100,000 swifts each summer in the UK, racing through the skies and giving joy to those who do notice them. 'The swifts are back, the globe's still working, our summer's all to come,' as Ted Hughes tells us. Our hearts rejoice at the return of every swift; so long as we catch that air-renting screech from a small number of birds, we fool ourselves into thinking everything is all right.

Yet the globe is not in good shape and most people know it. The wreckage wrought on the natural world is frightening, and because we tend to feel powerless to do anything about it, we sweep this disturbing knowledge into the dark recesses of our minds. I have spent twenty years working for a conservation charity, so the blinkers are off, I cannot escape the latest bleak statistics. I have grown a tough carapace, though, mental armour to protect myself from being dragged into a black hole of despair. A certain kind of mortuary humour pervades the office at times, but with gritted teeth you take the latest red listing of some unfortunate creature to try and unlock funds that might help do something to save its habitat. Often I skip the news stories about rainforest destruction and biodiversity extinction, pausing briefly to sign a petition against some global brand or other that threatens environmental havoc, vowing silently never to buy their products.

Fortunately, not everyone shuts their eyes to the fact that once-common species are failing to thrive in our fast-changing world and dwindling into rarity. Such people do not wait for official designations and conservation strategies to creak belatedly into action. There are those whose instinct is to do something themselves, who believe that change starts with local, personal actions, and that if there are others who share this outlook and make similar changes, the impact could be huge. 'Think global, act local' became a catchphrase for environmental action in the 1970s and its power could nowhere be better exemplified than in the grassroots movement for swifts, which began in the last decade of the twentieth century. Gradually, it has spread across the country, with parallel groups springing up around the world, growing collectively into an influential force.

One of the first people to notice that swifts were experiencing problems was a builder in Cambridgeshire called Bill Murrells. He had watched them on their return to their breeding places in the old quarter around Ely Cathedral and noticed they were unable to access their holes. The birds were banging their heads against the exact points where the entrances had been the year before but which had, over the winter months, been blocked with mortar during renovation work. Soon after, Bill read an article about swifts in his village newsletter and contacted the author to tell him what he had witnessed. So it was that he and Jake Allsop, who had just retired from a global career in educational training, got together. They realised straightaway that nest hole blocking was almost certainly occurring across the UK, with potentially disastrous implications for swifts.

There was another man in East Anglia who loved swifts, too: Chris Mead, the legendary media spokesman of the British Trust for Ornithology, a formidably knowledgeable man with a pyrotechnic way with ideas. The three of them formed a group called Concern for Swifts. A cautious sort of a name, perhaps, but they had bold ambitions. Their first thought was to do whatever they could to save existing colonies and prevent further destruction. To this end, they produced a leaflet giving simple information and diagrams showing how to make provision for breeding swifts while doing renovation work. What they had not anticipated was the way building regulations would constantly be revised, such as increasingly strict rules on insulation, making it virtually impossible to keep their recommendations up to date. Eventually, nest box installation became the priority, with conservation of existing nest accommodation a bonus where it could be achieved.

The group attracted architects, builders, bird ringers and other swift-watchers from East Anglia, sharing their knowledge and practical skills. Each year they would meet at Chris's home in Norfolk for a feast of bouillabaisse, and inspire each other to keep going. Jake remembers him with affection: 'Chris was a big man with a big beard and a stentorian voice. He had enormous energy and enthusiasm and was a born communicator.'

Meanwhile, in the north of the country, Concern for Swifts (Scotland) was set up by Clare Darlaston, inspired by the original group. In the peripheral estates of the 1950s, built around Glasgow, Motherwell and Dundee, swifts had found plenty of places to nest. Constructed before the invention of plastic soffits, gaps had opened out beneath the guttering and they bred in good numbers, delighting local people with their wild, aerial displays. Then a wave of renovation swept in; the holes were sealed and the birds lost. Scotland's swifts have undergone a 59 per cent decline since the 1990s.

Chris Mead died in 2003, leaving a gaping hole in the organisation he had powered into existence. Fortunately, while working on a barn owl project in the Fens the following year, Jake met another influential person: Dick Newell. He was the founder of Smallworld, a software company based in Cambridge that became the global market leader in geographical information systems (GIS) for utilities and communications companies. In 2000 he had sold it to General Electric and had semi-retired. Now, he is throwing his considerable energy into bird conservation. The entrepreneur who had once grappled with the complexities of creating enormous database systems for Deutsche Telecom and EDF Energy joined Jake and the others to reinvigorate Concern for Swifts, starting with a new name: Action for Swifts.

I take the train to Cambridgeshire to meet Dick. Speeding through mile after mile of hedgeless, weedless arable fields, I wonder where swifts might find food in such an intensively-farmed landscape. Mostly probably they fly to the Fens; the air over those ancient marshes should still be alive with insects. Crammed with tourists from China, Japan and America, the carriage empties at Cambridge and I change to the King's Lynn train on the Fen Line, alighting at Waterbeach, a station with no buildings and no staff but still a decent train service.

Dick lives with his wife Vida in a rambling, seventeenth-century farmhouse and has always been a keen birder. Previous obsessions have included skuas and gulls: 'We live between two rubbish dumps so there are plenty of gulls about! I've seen both Caspian and yellow-legged from my garden.' The day is warm and we are sitting outside. While he talks, Dick's ears are tuned into birdcalls: 'Siskin! Long-tailed tits! Tree-creeper!' Back in the Newells' kitchen, a TV screen in the corner shows a brood of downy barn owl chicks in a nest box just twenty yards from the house.

One summer's day Dick had noticed swifts flying up to the eaves of his home, searching for an entrance and somewhere to nest. He swung into action and fixed up some boxes for them. There are nineteen now, all of which he has made himself. Numbers fluctuate, but so far twelve of these have been occupied by swifts, with a maximum of nine in any one year. Cameras are fitted into some of the boxes so he can watch and record the progress of the breeding birds and their young. He has also put up a differently designed box for starlings, which can be a pest to swifts. In competition for nesting holes, they sometimes jump on their backs and push them to the ground. Starlings need help too, though; their numbers have declined by 66 per cent since the 1970s.

Ever the innovator, Dick likes to experiment and develop ideas. After a career spent at the cutting edge of digital technology, he works happily with wood, hammer and saw, experimenting with different designs of swift box to suit different locations and to try and find out which kind of nest box swifts prefer. Painting them white gives better protection from the sun, as does a double plywood roof. Might it help the birds if something nest-shaped was installed in the box? Given they spend a whole season gathering nesting materials in the air, so that the following year their eggs will not roll around, might this give them the chance to breed earlier? Dick has devised an easy way of making nest concaves out of fibreboard. Other options are lipped cork coasters or a simple square of wood with a circle cut out of the centre, fixed to the bottom of the box.

In St Mary's Church, Ely, he placed these concaves in every other nest box of the twenty-four installed. Ten boxes were occupied that year, seven in those with ready-made nests. This was encouraging, but not quite statistically significant. A further experiment at All Saints' Church, Worlington, in Suffolk, gave strong confirmation that swifts really do prefer them. Once again, half the boxes were fitted with concaves, half without. Of six boxes occupied, five had artificial nests. Further research has shown that, when given the choice, swifts are four times more likely to choose a box with a concave than one without. Birds do not give up their nest-making habits with these ready-made fixtures, however, continuing to add feathers and other airborne materials each year.

Attracting swifts to new nesting holes takes time. The

process is speeded up if they can be enticed with their own calls. Dick has spent hundreds of hours putting together sound systems, ingeniously assembled from a combination of golf cart batteries, solar panels, car tweeters and MP3 players. A breakthrough came when he discovered the Cheng Sheng player/amplifier – a low-cost device which could be controlled by a timer switch; a digital timer is preferred as it can survive power cuts. Simplicity is essential when trying to get people to do new things, and this system answers that perfectly.

With Dick at the helm, Action for Swifts has embraced the nest box as a solution to the loss of holes in old buildings and the dearth of them in new ones. Since making his own house swift-friendly, he has been involved in sixty-five other nest box schemes in the Cambridge area, putting up over 1,000 boxes in churches, schools, housing estates and under bridges. It was Dick who put up the boxes in Judith Wakelam's local church.

We take a tour of a handful of these projects. It is August and the skies have been growing quieter all week; we both think we have seen the last swifts of summer. Then, as we arrive at Edgecombe, near the Science Park, we hear a familiar scream. My heart leaps; a few late lingerers are still with us.

More than seventy wooden nest boxes have been installed on a block of flats, but most have been snapped up by house sparrows. Their nesting places are disappearing too, a contributory cause of their alarming decline over recent decades. One resident decided to install a sound system to play swift calls and his boxes quickly attracted swifts. It is easy to guess which flat is his. Peter Glass's first-floor balcony not only has swift boxes but also a hanging bird feeder filled with peanuts and two tit boxes fixed to the walls. 'I was sitting out with a cup of tea when the young blue tits left the nest box, one by one,' he says, beaming with pleasure.

In a study of swift nest box projects in 2016, a researcher for the RSPB spent many hours watching the Edgecombe flats, noting twelve boxes occupied by swifts, and many more with evidence of house sparrows. Some of the boxes had previously been occupied by sparrows; the swifts had simply added their own materials on top of the house sparrow haystacks. Dick hopes the success of this project will encourage other councils to replicate the experiment elsewhere. Swift boxes are greatly preferred by sparrows to the much marketed sparrow terrace boxes. Like swifts, they prefer to nest in loose colonies and not too close to each other.

While swifts do not generally take to boxes as easily and quickly as a blue tit or a house sparrow, determination, experimentation and persistence are bringing results. In the village where Dick lives there were four breeding pairs when he first started looking out for them. Two of the houses where they nested were subsequently demolished and rebuilt, without any attempt to provide new holes for the birds. Instead of being overwhelmed with fury at the indifference of developers to wild birds, he explored alternative possibilities and put up more boxes. There are now some thirty breeding pairs in the village.

A contemporary concept for the provision of nesting habitats that is taking off all around Europe is the swift tower. There are two of these in Cambridge, the first of which was built in Logan's Meadow, an urban nature reserve next to the River Cam. With funding from section 106 levies paid by housing developers to the council, an art installation was created incorporating 100 swift boxes and eleven bat boxes. Erected on a pylon-like steel structure, the huge disc of boxes looks like a circular Rubik's cube; it is painted in vibrant reds, oranges and purples to evoke the idea of the setting African sun.

Predictably, this £35,000 art project caused a bit of a

Swift tower on Logan's Meadow, Cambridge,
designed by Andrew Merritt.

storm in the *Daily Mail*, which castigated 'Britain's most expensive bird box'. Its first occupants were starlings, which somehow managed to squeeze themselves into the tight little holes designed for swifts. The council, with help from Action for Swifts, reduced the size of the entrance holes and swifts started to investigate the tower, attracted by calls played by a modified bird scarer (now replaced with the magical Cheng Sheng). Wild screaming displays took place around it, and in 2015 a pair was seen entering a box. Dick tells me he would often take a chair and sit in Logan's Meadow, waiting for hours to see if swifts were using the boxes. On this warm summer day, we stroll down to the tower and instantly a swift peeps out of one of the upper holes and flies off. Such luck!

Despite the huge price tag, he reckons the tower is valuable. It is a wonderfully eye-catching structure and many people stop to look and ask what its purpose is. Occupancy rates may grow. Standing on the footbridge in June and July, you can watch swifts on summer evenings, a perfect place for people to enjoy and become acquainted with these awesome birds.

The Trumpington Community Orchard swift tower is quite different. Here, in a village on the outskirts of Cambridge, heritage fruit trees have been planted next to some allotments, with knapweed and field scabious in the grass to keep bees and butterflies supplied with late-summer nectar. The chair of the orchard committee had suggested to Dick that this would be a good place for a swift tower – so he offered to make one for a fraction of the cost of the Logan's Meadow installation. It is very

Raising the tower at Trumpington Community Orchard.

much like a barn owl box, a steep-sided triangular design with eleven compartments for swifts.

The box and its eight-metre telegraph pole were winched into place with a tug-of-war team of helpers using a specially constructed A-frame of scaffolding, on which to rest the telegraph pole between lifts. For good measure, two bat boxes and a squirrel barrier were fitted to it, along with a solar panel to keep the call system battery energised. In its second summer, up to twenty screaming swifts were seen overhead and, even more thrillingly, a pair nested in one of the boxes, with the quiet clamour of hungry chicks heard when the parents delivered food. It is exciting when something as simple as putting up a box with a hole in it has such positive results for creatures you care about.

Swift towers have not, however, been as successful as hoped. Installed in a number of places across the UK and the continent, with the promise of an instant solution to the loss of nest sites in traditional places, swifts have not shown as much interest in them as some people expected and occupancy rates are low.

A simple box or two under the eaves works much better. Inspired by the 'just try it' attitude of Action for Swifts, Rowena and Clive Baxter of Dry Drayton decided to do just that when they saw swifts flying to the eaves of neighbouring houses in a cul-de-sac built in the 1970s. At the time there were no known breeding colonies locally, but they installed boxes on their house and played swift calls to entice them. This involved climbing in and out of a bedroom window to switch the sound system on and off. Then they persuaded their neighbours to do the same thing. Better technology had arrived by then, so it was no longer a perilous operation. Five years on and there were twenty-three nest boxes on eight houses in one street, with twenty chicks reared.

The swift conservation dynamic is a powerful force, thanks to the work of this group. All the projects undertaken are written up in detail on the Action for Swifts blog, and enquiries are dealt with quickly and helpfully. Excitement sparks from them, inspiring many other people to set up their own nest box schemes. There are now ninety groups around the UK taking action for their local swifts. 'It's dead simple to help swifts – all it requires is someone with a screwdriver, a saw and a tape measure to get on and do it,' says Dick with conviction. Some 350 pairs of swifts nest in the boxes he and others have put up in Cambridgeshire over the last few years, living proof of the effectiveness of this strategy.

Dick has also taken a leading role in migration tracking projects. Approached by the British Trust for Ornithology in 2010, he agreed to help with the purchase of light-logging devices for swifts, to enable the first migration tracking project in the UK. European research scientists had already started to make use of new technology to learn about the journeys of swifts, and ornithologists in the UK were keen to have a go themselves. Ten tiny data loggers were bought and, at the end of the summer, one was fixed to a swift nesting on Dick's house.

'When the swift returned the following May, it was a terrific moment!' he says. The bird had brought with it a complete record of its journey. Much of what we refer to as its 'wintering period' had been spent in the summer skies over Mozambique, wandering constantly, before moving northwards on its return migration, leaving the Congo region on 20 April and arriving back in its nest box on 12 May. Along with others in the tracking project, it took a detour westwards, lingering over Liberia for ten days. The pioneering project initiated by Swedish scientists in 2009 had first revealed this detail in the swift's migration. From there, Dick's bird took less than a week to complete its 3,000-mile journey, travelling an average of 435 miles per day. The final leg, from northern Morocco back to Cambridgeshire, a distance of over 1,000 miles, was completed in just two days.

He was involved, too, in the Beijing swift-tracking

project. For some years he had led bird-watching tours there, working holidays away from the massive data system development projects in Cambridge. During these expeditions he had developed a network of birding contacts. In partnership with them he initiated the tracking scheme, bringing international swift researchers together with ornithologists in China. And revelations about migration routes have not been the only outcome: inspired by the work of Action for Swifts in the UK, Beijing residents have begun to install nest boxes. It has become a global movement.

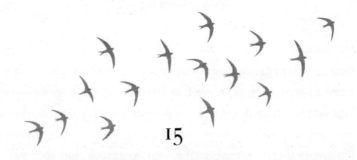

15

GLOBAL ADVOCATE

Another powerful voice for swifts emerged in London. Edward Mayer was first attracted to these birds at the age of six, when they screamed around his home in Southampton. People who start looking at birds in childhood never forget them. Living in north-west London for the last fifty years, swifts have been regular summer companions for Edward, watched for and welcomed back every May.

For much of his working life, Edward managed the Tate Gallery buildings, including storage sites and archive repositories. He was well aware that urban architecture provided good nesting opportunities for a range of birds and never saw this as a problem. Indeed, it brought some very obvious benefits. On one occasion he was asked to sign a contract with a vermin control company, to agree that rodenticide should be regularly put down around the Tate buildings to exterminate rats. He refused. 'I had never seen a rat around and knew that a pair of kestrels nested on the Millbank tower – they would take care of any rodents.'

Edward Mayer (right).

In 2002 he was dismayed to notice the roof of a council-owned property near his home being removed for renovation; he had watched swifts entering the eaves of the building for years. In time-honoured fashion he fired off an outraged missive to his local paper and the editor featured it as the week's star letter. Next, a reporter rang him and wrote a half-page feature about swifts and the problems they were facing. Then Camden Council invited him to join its biodiversity committee. His intervention had struck a chord with people: something had to be done to give the local swifts a chance.

Seeking advice on how to spread the word, he contacted the celebrated swift guru, Chris Mead. 'You must have a website,' thundered Chris. And so it was that Edward grappled with new technology and launched London's Swifts, offering advice on how to help the capital's swift population. At the time, there was little information on the web about the problems they encountered and how to resolve them, and his first big emergency came not from London, but from a town called Alcúdia, near Valencia in Spain, where a whole complex of historic buildings associated with the church was being restored. No environmental impact assessment had been carried out and the architect had inspected the buildings only in winter. Now it was spring and all the walls had been netted; swifts, swallows and redstarts had returned and were becoming entangled or crashing to the ground, unable to access their nest sites – the identical problem to that which had occurred in Ely.

Thankfully, once the authorities in Alcúdia realised what they had done and its disastrous effect on the birds, they did everything they could to put things right, recognising that swifts and other nesting birds were an integral part of the interest and character of the town's historic buildings and that this wildlife was something to be celebrated. Conflict resolution does not often come this easily.

Next, Edward took on a local project at London Zoo, where measures were underway to encourage more native birds to breed. Swift boxes were put up in the bear caves and under the eaves of the insect house. Swifts moved in and bred for several years, but disappeared again after a series of cold, wet summers.

Edward has persuaded people to make space for swifts on scores of private and public buildings in London and beyond, including Canada Tower, New Scotland Yard, housing developments in Stepney and Dulwich, a synagogue near Golders Green, office blocks and several prisons. 'We didn't have funding for many of these projects but local councils gave us ghetto blasters they'd confiscated from raves, so we could play the swift calls!'

Not everywhere is suitable. Glass buildings are hazardous for birds as they can crash into their reflective surfaces. 'Bird kill is high in the glass canyons of the city; only by the riverside would I put up swift boxes. Where whole areas are being redeveloped, or where flood-lighting and laser beams are present, there is probably too much disturbance for the birds to try and breed.' Edward is currently campaigning for swift boxes to be incorporated into the Houses of Parliament during the planned renovation project. Curiously, these buildings were fitted with one of the earliest air-conditioning systems, to clean soot from the filthy London air. This drastically reduced the need for ventilation holes, making the buildings inhospitable to swifts. Now, there is a chance that purpose-made nesting holes can be integrated into the building, giving the birds a future in the heart of the capital.

London's wildlife habitats are also deteriorating in the very place you would think they would be safe – its gardens. The contemporary fashion for designing gardens as minimalist outdoor rooms has encouraged a great deal of paving instead of planting; nearly two-thirds of all London's front gardens are now covered with concrete or stone slabs. According to a survey by London Wildlife Trust, within an eight-year period at the turn of the century an area of vegetated gardens twenty-one times the size of Hyde Park disappeared under hard surfacing. This impoverished environment withers people's spirits and seriously depletes the city's invertebrate population, making it harder for insectivorous birds and mammals such as hedgehogs to find food. Swift numbers in the capital have halved over the last twenty years.

While Edward and others are bringing sparks of enlightenment to every local planning authority that will listen, the elimination of cavities during roof restoration continues across the UK. It often comes in waves. A massive government spending programme to modernise schools, hospitals and housing in Scotland at the beginning of the century was essentially a good thing – except that it took no account of the birds that had nested under the eaves of the roofs, sometimes for 100 years. The effect on Scotland's breeding swifts has been devastating.

Schemes are underway to address the loss of nest sites in parts of the country, including Edinburgh and Tayside, with developers encouraged to include nesting bricks in new buildings and fit nest boxes to old ones. It will, however, take years of concerted activity to reverse the losses and restore the great midge-eaters to their former numbers.

Meanwhile, another wave of change is looming over

The swift's feathers shimmer with iridescence as it emerges from a nest brick into sunlight. (© Simon Stirrup)

Parent swifts collect between 300 and 1,000 insects per meal for their chicks. (© Stephen Barlow)

The aerodynamics of flight. **Opposite** A steep, banking turn, the wings almost vertical. **Above** Tilting for a turn. **Below** A slow glide, the wings slightly turned down. (© Piotra Szczypa)

Above Swifts at sunset. (© Ben Andrew)

Opposite The feathers of young swifts have white edges and the throat is paler than in the adult. (© Ben Andrew)

Below Common swift drinking from a forest lake in Denmark.
(© AGAMI Photo Agency/Alamy Stock Photo)

Aerial mating of common swifts. (© John Hawkins)

The swift's throat is engorged with food for its nestlings. (© Roger Wyatt)

Swift chicks at 35, 41 and 42 days old. (© Ulrich Tigges)

Swifts flying around Gaudi's
La Pedrera building in
Barcelona. (© Laurent Godel)

Reading in Berkshire. 'Crossrail is coming and this will almost certainly lead to an influx of commuters and renovation of the nineteenth-century terraces,' says Edward, his voice full of dread. It is always possible to incorporate holes into such projects – to do away with rot and decay while keeping access for swifts – however, it is unrealistic to expect that this will generally happen unless a swift-aware roofer happens to be undertaking the job. So Edward is taking a pragmatic approach and working with the local Catholic church to install swift boxes and a call system in the bell tower, ahead of the inevitable losses. He hopes that this project will inspire others to take action and replicate the idea elsewhere in the area.

Edward is a man of persuasive charm and exuberance, and you can see exactly why swifts, with their exhilarating lives, appeal to him so strongly. And why he is so good at getting people to listen. He has given hundreds of talks, lectures and seminars, inspiring people to take notice of swifts, and do something for them. Today he often finds himself addressing architects and local councils, raising awareness of how swifts can be accommodated in modern buildings. His professional background in the maintenance of the Tate estate equips him well to discuss the practicalities of adding holes to buildings, something many architects are naturally suspicious of. 'I like to show them how swifts enhance the environment and the drama of the building – this catches their imaginations,' he says. Architects in the UK are not taught about wildlife during their training, unlike in Switzerland, for example. A green roof in Zurich will be constructed using only materials and plant species occurring within the local canton, integrating it into its environment perfectly. In the UK the materials may be sourced from anywhere in the world, often rendering them inhospitable to native flora.

Edward's energy and zest for his subject make him a sought-after and influential speaker. Showing a photo of brutalist, concrete desolation, he exclaims: 'I too would have a pitbull terrier and terrorise passers-by if I lived in this grim environment!' He is scathing about many architects' lollipop tree approach to landscaping, advocating instead a vision of the Hanging Gardens of Babylon, with ivy and other vegetation clambering up buildings. You come away from his presentations with a real sense of wonder; his passion for swifts is infectious. The substance of his advice, however, is always anchored in practicality. He shows bold, beautiful and simple projects from around the world that demonstrate what can be done to give birds an even chance in our cities, while also providing people with attractive surroundings.

From a local start in London, Edward's conservation activities have expanded to a global level, working for swift conservation throughout the world and bringing together swift experts and local activists through conferences and training sessions. London's Swifts soon morphed into the all-embracing Swift Conservation, a central point for enquiries about swifts and how to help them. He has developed a vast network of swift contacts around the world. A fluent French speaker with a smattering of Italian, Spanish and German, he is at ease internationally and works regularly in Italy and France, providing training in supporting and enhancing urban biodiversity to architects and local government staff, as well as helping with popular events, such as swift walks and evenings, spotting swifts with a glass of wine in one hand, binoculars in the other.

Sharing ideas and experiences is the best way to motivate people and create momentum for change. Wondrous though the internet is for finding out what is going on around the world, it is no substitute for meeting people and sensing the excitement and passion for these extraordinary birds. In 2010 Edward helped arrange a conference for swift conservationists in Berlin. People arrived from Poland, Holland, Germany, Switzerland and the UK. Two years later, at the second conference, they were joined by China, Russia, Romania, Belgium, the Czech Republic, Slovenia, Turkey and Israel. Four years later, the conference was held in Cambridge, by which time Uzbekistan, Azerbaijan, Brazil, Italy, the USA and Canada had got involved too. 'A hundred and fifty people from twenty-one countries attended that conference, every one of them enthusiastic and dedicated, there was a real buzz and the talks, given by swift experts from around the world, were wonderful,' says Jake Allsop.

Yet apocalyptic gloom affects even Edward sometimes: 'If we can wake up from the dangerous internal world we are rapidly heading into, where nearly all our waking hours are spent staring at electronic screens, and start to realise that the natural world out there is the one that will keep us sane, level-headed, focused and happy, then we will ensure that swifts will be flying above our heads as we step into the future. Otherwise I fear we all may be doomed to increasing isolation, widening paranoia and eventually a form of societal madness.'

Despite this, he has a strong belief in the ability of people to bring positive change: 'Almost invariably it has been individuals and not institutions that have advanced swift research and supported and enhanced swift breeding. Look at the swift groups in Northern Ireland and Frankfurt, Action for Swifts in East Anglia and Stephen Fitt in Exeter!* At a personal level, they have brought about the installation of thousands of nest boxes far and wide.'

* See Chapter 17.

16

SCOTLAND'S SWIFTS

The decline of swifts in Scotland over the last twenty years has been particularly acute. Clare Darlaston of Concern for Swifts (Scotland) witnessed the change:

Ten years ago I used to walk through streets in parts of Glasgow in summer and see and hear screaming parties displaying around the buildings until nearly dark-time, after 11pm in June. It was never as exciting as, for example, Castle Douglas, where the buildings are lower and the swifts [sometimes] scream down the street at knee level, but it was the ever-present sound of their calls on fine summer evenings that alerted me to their presence and their thrilling chases around the buildings. Now I wish I had paid more attention to recording numbers, as year on year they have become fewer, even in areas not badly affected by renovation.

Her perception is shared by Roy Dennis, pioneer of osprey reintroduction schemes, who has lived and worked in the Highlands of Scotland for sixty years. Roy had a distinguished career at the RSPB, after which he set up his own wildlife foundation, which gives him the freedom to concentrate on species recovery projects and the restoration of natural ecosystems. He too has witnessed the decline of swifts and talks wistfully of the days when you could watch hundreds of them flying around Forres Castle, close to his home in Moray. 'Today, you are lucky if you see ten. A big change came in the 1980s and 90s, when farmers began to spray their crops intensively – that's when the numbers started to decrease. Fifty per cent of land should be left to nature; every farm should have wildflowers,' he says, speaking with passion.

Academic studies demonstrate that the abundance of insects in Scotland has dramatically diminished. Research carried out by the University of Stirling in 2002 showed an average decline of 49 per cent in the fifteen major groups of the swifts' prey between 1983 and 1997.

Daniele Muir worked as a ranger with Perthshire Council for fifteen years, with special responsibility for swift conservation. When I ask her why she thinks the decline has been particularly severe in Scotland, she suggests that it may simply have been the case that change came later and faster than in England, causing a more sudden downturn in numbers. 'Building renovation, new developments and dwindling insect populations – the same negative factors are present here as everywhere else,' she says.

During her time at Perthshire Council she helped set up many nest box schemes on schools, churches and houses. She grew downhearted at the inadequacies of the planning system in supporting the integration of nest bricks into new buildings. Housing development is booming in Scotland but very few new buildings have swift bricks, despite the recommendations of the RSPB and often the various councils' own biodiversity officers. Even when such recommendations are included in planning conditions, they are seldom implemented.

She now runs her own business, leading wildlife walks in the Perthshire hills and kayaking along the River Tay to show people beavers. It might seem that Daniele has a dream job, but she would give it all up to work solely for swifts. 'I would love to be paid to work for swifts full time. They are such captivating birds – people fall in love with them and will take action to help them when they know their problems, but so much more needs to be done. We have to get the message across to many more people to bring change on a scale that really helps them.'

It is one of the ancient delights of swifts that people can enjoy them wherever they live, in cities, towns, villages and deep in rural country. Set on a hill west of Inverness, overlooking the meanders of the River Beauly, is the House of Aigas. This pink sandstone, neo-Jacobean mansion, ornamented with towers and turrets, was built 'to prop up the spiralling self-importance of the wealthy Victorian gentry'. These are the words of Sir John Lister-Kaye, who was looking for somewhere that would double as a family home and a field studies centre. He had escaped a career in the steel industry and spent time with Gavin Maxwell on the west coast of Scotland, soon realising that he wanted to stay in the Highlands and write books about natural history. In his first glimpse of Aigas in the summer of 1976, his heart was caught by the sight of swifts hawking and screeching around the tower.

Undeterred by damage wrought on the house by decades of neglect and decay, he was determined to buy it. That winter he finally set foot inside, finding heaps of shattered glass in the hall and a snowdrift at the bottom of the stairs.

Rescued from the brink of demolition, the house is now a warm, welcoming place, where people, young and old, come to find out about nature. They stumble delightedly on beaver-chewed wood around the lake, watch pine martens at dusk, wonder at glistening sundews and golden-ringed dragonflies. People book time here to watch eagles and harriers or to study tiny invertebrates, to take excursions into the hills to see red deer or enjoy walks along valleys looking at wildlife. And when they get back to the turreted house on summer evenings, they will still see swifts racing around the rooftops.

When he first moved into the house, Sir John would sit in the roof space of his tower watching a pair of nesting swifts, just as David Lack had done in the Oxford Natural History Museum. He offers to show me this attic eyrie, first engaging the assistance of his two work-placement wildlife rangers, Jess and Jay. Ladders are off-limits for him now – he is awaiting a knee replacement – and with wistful resignation he stays put at its foot. A large scattering of filth falls to the floor as Jay opens the hatch and we climb up.

Sir John's old, striped garden chair is slung across the beam in the apex of the roof, dimly illuminated by light filtering through the louvres of the narrow air vents. 'It was slightly spooky watching the swifts, eyeball to eyeball,' he remembers. When the nestlings were close to fledging they would sometimes crawl up the walls, resting vertically. Wrapped in darkness, it would have been an intimate experience, quite different from the contemporary method of watching them on camera. He has a camera now, though, with a screen in the common room, enabling all the Aigas visitors to catch a glimpse of the swifts preening, fidgeting, sleeping or occasionally feeding their chicks.

This year the swifts are absent from this nesting site, but an enormous heap of dried grass, some eighteen inches high and two feet wide, bears testament to their occupancy over many generations. During the breeding period they continuously bring in nest material, a lot of which has been dropped, creating a small mountain of debris. Mouse burrows show that it has been put to good purpose.

We look in the other section of the partitioned roof and see the swift that was on livestream video link below. It is good to share a few minutes of its company in that dusky attic.

Two swift nest boxes had been installed the previous year and are now occupied, along with another pair in a crack in the wall. Swifts are doing better at Aigas than house martins and swallows. When Sir John moved to the house in 1976, a dozen pairs of swallows were nesting around the buildings and he also noted twelve or more active martin nests. Numbers dwindled steadily over the years, and by 2019 there was just a single breeding pair of martins; the swallows had gone. Flocks of hirundines were seen hunting over the fields of the estate in late summer; perhaps they will be back.

An air of melancholy comes over Sir John when he talks about the problems faced by insectivorous birds. He is deeply concerned about the decline in invertebrate abundance caused by changes in land use and widespread use of pesticides, worrying about the situation here and across the world.

There is one place in Scotland where swifts seem to be entirely unperturbed by the problems engulfing their species elsewhere. Their numbers are few but they are living in much the same way they might have done thousands of years ago. This is Abernethy Forest, just south of Inverness.

Like huge cushions of moss, the emerald leaves of bilberry bushes light up the ground beneath the trees. Uwe Stoneman, conservation manager of the RSPB's Abernethy Forest nature reserve, loves this part of the woods. This is where the oldest Caledonian pine trees can be seen, the grey of their deeply fissured bark grown misty-green with lichen. One fire-scarred pine dates back to 1640, but the forest as a whole is vastly more ancient; this tree has direct ancestry with the Scots pines that first grew across the Highlands at the end of the last Ice Age, 9,000 years ago. About eighty scattered remnants of

Caledonian forest survive today, many of them tiny. Abernethy is one of the largest, covering over 3,000 acres.

The RSPB is also a partner in a landscape-scale project with a 200-year vision, Cairngorms Connect, which aims to double the amount of pine woodland on the Abernethy national nature reserve, alongside a programme of ditch-blocking on neighbouring moors to rewet peat bogs, locking in carbon and providing the soaking conditions that sphagnum mosses, bog asphodel and cotton grasses require.

All along the track young pine trees are sprouting up, growing more prolifically here than deeper in the woods, where less sunlight falls. It seems the natural thing, to see the next generation springing up, but for over half a century young growth was rarely to be found. The contin-uous munching of a vastly expanded deer population on the tender pine needles and juicy young leaves of aspen and rowan seedlings effectively suppressed new growth.

Keeping the deer population in ecological balance is one of Uwe's main tasks. This is necessary not just for the future of the woods themselves but for the wildlife that lives in them, and in particular for capercaillie, the world's largest grouse. The male of this species is turkey-sized, gorgeously feathered and with a display call like popping corks. Their numbers dwindled in the eighteenth century, largely because of loss of their woodland habitat, but also as a result of hunting.

Victorian landowners wanted this ultimate game bird back on their estates and brought some over from Sweden. For a while they flourished, but in the 1970s their numbers

started to tumble. Climate change has brought frequent
wet spring and summer weather and resulted in poor
breeding success, while the deer fencing erected over
hundreds of miles of the Scottish Highlands has been
calamitous for both capercaillie and black grouse, with
fatal collisions common. There are now thought to be as
few as a thousand left, leaving them once again on the
brink of extinction in the UK.

However, it is not the beleaguered capercaillie that
brings me to Abernethy, nor the Loch Garten ospreys –
for the first time in a century they have failed to return
to their historic breeding site – but something else vanishingly rare. This is the only place in Britain where swifts
are known to nest in trees. Old woodpecker holes are one
of the few natural nesting places used by swifts but would
once have been their most available option. Tree-nesting
swifts are still to be found in old-growth forests in Eastern
Europe and Finland, while one was recently filmed in
Barcelona, entering and leaving a hole in a pollarded plane
tree. It would be an awesome experience to see them in
these ancient Scottish woods, where they have perhaps
nested for several thousand years.

Swifts, Uwe tells me, are his wife's favourite birds.
However, he has only recently taken up his role at
Abernethy and not yet seen the tree-nesters. So he has
made enquiries among his colleagues and been directed
to a large, dead tree with a great many juniper bushes
nearby.

We leave the track, treading our way through bouncy
heather and bilberry. It feels strange to be looking for

swift nests away from the built environment, as though we are stepping back into a vanished age. Quite soon we come across a gaunt, silvery old pine amid the sprawling branches of juniper. The tree does not, however, have any obvious woodpecker holes. We hang around for a while, slapping our heads constantly to rid ourselves of the persistent midges.

Uwe strides off one way and I go another, to see if there are any other dead pines nearby. Then Uwe shouts out – he has found the right tree and seen a swift fly out of a hole. This tree is much taller, a mighty skeleton, higher than the living pines that touch its claw-like branches. We settle down at the mossy foot of another tree to wait for its return. If the swifts are incubating, it might be that Uwe witnessed a changeover of the parents and it could be many hours before there is any more activity. Quite possibly, though, it was delivering food to nestlings, in which case it will return with another ball of insects in about an hour.

Eyes fixed high on the tree, on the hole from which the swift had flown, I cannot look away for a single second in case it returns. The forest is surprisingly quiet, the only bird calling a solitary chiffchaff. 'A few weeks ago there would have been wood warblers singing from every tree!' Uwe says. It is quite something, this silence. If it were not for the midges it would be utterly tranquil, but suddenly I feel as if I will go mad with their insistent bothering and biting. I am ready to go; it is enough that I have seen a tree where I know that swifts nest, seen the woods where the fledglings will take their first flight.

Scots pine in the Abernethy Forest where
swifts nest in old woodpecker holes.

Then, into a scrap of sky between the treetops glides
a swift. And then another. Three of them! One is calling.
In broad circuits they cruise around the old tree for a
minute or two. And then one disappears in a flicker of
feathers into a hole above the one we have been watching;
another bird follows a second later, slipping into the round
cavity our eyes have been fixed on. Fortune has been kind;
another minute of waiting and my impatience would have
got the better of me and I would have missed this.

Uwe is glad to have seen this, too. There are three trees

in the Abernethy area where swifts are known to nest, but he suspects there may be more, here and elsewhere. Numbers could, however, be limited by the relative lack of dead trees in the forest. Timber production over the centuries has impoverished the age structure of the forest, something the RSPB made an imaginative attempt to remedy a few years ago when they detonated explosive charges on ten living trees to simulate the effects of lightning and wind damage, so the trees broke off in a ragged way.

Inevitably, there was a backlash against a conservation charity blowing up trees, but for the forest as a whole it was a life-enhancing strategy, providing habitat for wood-boring insects, mosses and lichens, and new cavities for bats and tree-hole-nesting birds such as crossbills and woodpeckers. With a 200-year management vision that promotes natural processes, time will eventually restore the dynamic of a diverse, thriving forest.

17

NEW HOUSES, NEW HOLES

The approach of the grassroots swift conservation movement works through inspiration, enthusiasm and a strong element of DIY. Motivated people fit boxes onto their homes and other buildings, wherever opportunities arise: direct action by individual citizens to change their local environment.

Across south-west England, the scarcity of holes in buildings is being tackled in a different way. While the practice of 'retro fitting' – fixing boxes onto existing buildings – goes on among swift aficionados, another strategy is being rolled out. Planners and developers are gradually being nudged into installing swift bricks into new buildings. More than a million new houses could be built in the UK in the next decade. While this is an alarming prospect for many people who fear the loss of green space, it is also potentially a great opportunity for designing and building urban settlements that integrate nature into the human environment. The person who has so effectively promoted this approach is Stephen Fitt.

I meet him at his home near Exeter, a converted farm building, surrounded by a large wildlife-friendly garden, where he tells me of his first encounters with swifts. As a twelve-year-old he had watched them fly into their nest over one of the windows of his classroom. 'I did little maths in the summer term,' he says. It was not until he retired that he had time to think deeply about swifts again. Twenty years later, while living in one of the older areas of Exeter, he found a grounded swift and picked it up. It was infested with *Crataerina* parasites and he dropped it. Fortunately, it flew off.

Ill-health struck him, causing severe mobility problems, so a physically active retirement became impossible. Despite this, Stephen has used his time wonderfully. After signing up as a volunteer with the RSPB he worked initially with one of their conservation officers in Dorset, reviewing planning applications in and around Poole Harbour. One proposal involved demolishing a hotel on the seafront and replacing it with flats. He suggested that swift bricks should be integrated, and the RSPB's South West Regional Swift Project was born.

He now works with planners, architects and developers, chiefly in the south-west, trying to bring about a shift in building design which will make houses, industrial buildings and other constructions better able to withstand the effects of climate change, better for people living and working in them and better for wildlife. 'I don't do single species conservation,' he says forcefully. 'We need to develop environments that favour us all.' Swift boxes, however, will benefit a range of wildlife in search of somewhere to nest, roost or hibernate.

As a volunteer with the RSPB, he has carved out an influential role, for which he has been recognised with their President's Award. 'I have made a great nuisance of myself,' he grins. Local planning authorities are legally required to consult conservation organisations before reaching a decision on relevant planning applications, making his contributions more powerful than they would be if submitted simply as an individual. Confrontation is not his way, however. He takes a co-operative approach, working with many different people to find common ground. 'I think my background in financial services gave me useful skills – it was my job to investigate people's wants, needs and ambitions and to come up with plans that suited their requirements, so I learned how to listen and negotiate.'

It is people who shape our world, for better or worse, and, despite the common perception that conservationists spend their days out in the field listening to skylarks and helping injured wildlife, this is generally not the case. Decisions about how our land is used, where development will be accommodated and in what shape, are made in offices and committee meetings. Only in the most enlightened organisations – or during a pandemic – do they occur outside.

Stephen worked closely with Exeter City Council on the biodiversity section of the Residential Design

Supplementary Planning Document (SPD). You are very likely thinking this sounds mind-numbingly dull – planning documents are not renowned for their inspirational qualities – but this one is different, developing a visionary approach to the housing needs of the future and earning the council an urban design award from the Landscape Institute. Exeter has a strategy to build 12,000 houses by 2026, some 600 each year. This scale of building demands careful planning – an opportunity that the people who wrote the SPD appear to have relished.

Attractive homes that are a pleasure to live in, with beautiful streets, gardens and open spaces, should be created. Housing design should consider views, links to the wider landscape; the nature of the landform, and its vegetation, biodiversity and wildlife habitats.

The document advises that:

new housing should be designed within its surroundings, so it connects to local facilities and encourages sustainable transport. The addition of trees and other natural features are recommended too; their presence in the urban landscape brings joy to people, and roosting, nesting, perching and feeding opportunities for wildlife. Their leaves absorb nitrogen oxide and other gaseous vehicle emissions, while also doing their bit to help reduce carbon dioxide levels through photosynthesis.

The landscape framework for a site must ensure that the various components of landscape, such as public open space, play areas, woodland, hedgerows, wildlife habitats and green lanes are well connected to each

other; site development should contribute to the green
infrastructure of the city and underpin the landscape
of the site.

The language may be a little unexciting (this *is* a planning
document), but the holistic approach to the built and the
natural environment is spot on.

So what exactly is green infrastructure? The concept
originated in the United States in the 1990s, an ugly but
useful term in planning-speak to describe natural features
that help solve urban and climatic challenges such as
storm-water management, leading to the equally dismally
named Sustainable Urban Drainage Schemes (SUDS),
reduction of heat stress through tree planting, better air
quality, and leafy, green, flower-filled places that are great
for bees and birds and contribute enormously to people's
happiness.

There are some beautifully simple solutions to urban
pollution. Swales are one of them: broad, shallow vegetated hollows that can absorb road-runoff, filter pollution
and give rainwater somewhere to soak into the ground,
rather than deluging the drains. Rain gardens are another
SUDS idea. These sound quite romantic in an English sort
of way, but they are basically small hollows, planted with
ferns and other plants that tolerate or prefer a good
soaking, into which rainwater from paved areas or down-
pipes can be directed.

But what have SUDS and SPDs to do with that sublime
bird, the swift? The answer is that it is all about integrating
nature into our towns and cities, making sure rivers and

streams are clean so that they have good insect life, making room for trees that beetles and bugs feed on, some of which will rise into the air and get snapped up by insect-ivorous birds. And what is good for birds tends to be good for people, too.

Stephen has made one very direct contribution to helping swifts and other birds that nest in buildings. Detailed, practical recommendations for the provision of nesting fixtures for swallows, house martins, swifts and bats are set out in the document, and the SPD calls for swift bricks to be incorporated into all new buildings, with a minimum of one per residential unit.

These bricks, which blend in easily among their fellow conventional bricks, have a hole at the front and a box behind. They are maintenance-free and do not require cleaning out. Swift bricks are a neat solution to an urgent problem, yet some property developers have voiced concerns that prospective buyers might object to buying a house with a nesting cavity, a view somewhat bizarrely endorsed by the writer of an article in *The Sunday Telegraph* in 2017, which began with the sentence: 'Most homeowners would despair at the prospect of swifts roosting in the eaves or house martins building muddy nests on their soffits.'

Inevitably, there are people who object to sharing their homes with any kind of wildlife, mostly because they fear the mess of their excrement. It is often said of swifts that they do not let their droppings fall outside the nest and that they are cleaner than the hirundines. This is largely true; they leave vanishingly few signs of their occupancy.

However, using this argument to persuade people to make room for swifts validates the idea that it is alright to object to other, messier birds around our homes. Swifts good! House martins bad!

Defecation is, need I say it, a universal function of living creatures. While it can be slightly annoying that martins leave clumpy white trails down the side of your house and perhaps splatter your car, this is easily dealt with by putting up a shelf under their nests and collecting the droppings for your compost heap. These birds take so little from our world and leave minimal trace of their presence – unlike *homo sapiens*. They fly thousands of miles to our shores, eluding storms and Mediterranean bird catchers to nest and raise their young, bringing us months of pleasure if we are fortunate enough to have them nest above our windows. Shockingly, some people still knock their nests down, attempting to justify such vandalism by saying that martins have fleas. Well, yes, they sometimes do, but these parasites are specifically adapted to their host and have no interest in human blood.

Thankfully, a survey carried out at Gloucester University into the attitudes of people living in houses with integral swift or bat bricks has shown that people are happy to accommodate them, while some would be positively encouraged to purchase a house that included such a feature. The swift brick attracts many different occupants and may also be used by sparrows, starlings, blue tits, great tits, hibernating tortoiseshell butterflies or nesting bees. These bricks are an invaluable twenty-first-century ingredient in new buildings, helping to ensure our future co-existence with swifts and other wildlife.

Cost is inevitably a factor in a developer's willingness to integrate swift bricks into their buildings. Until recently, a good swift brick retailed at around £100 – a big price for a hole. Now, though, the price of these special bricks is coming down dramatically, while their design is getting better.

So, is Exeter's beautiful masterplan being turned into reality? This, of course, is the mountainously difficult bit. The SPD is a guidance document, and, while some of its advice has statutory clout, much of it does not. Even since its publication, the legal framework has been weakened: the policy for all new housing to be zero carbon, introduced by the Labour Government in 2006, was regrettably dropped by the Conservatives a few years later. Deep cuts in local authority spending have also led to a reduction

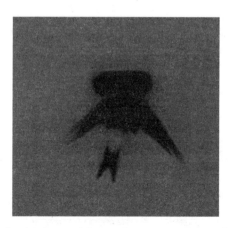

Nesting bricks designed for swifts attract a range of wildlife including, as above, house martins. Eleven pairs are known to have used them in a housing development in Truro, Cornwall.

of biodiversity staff at many councils and a weaker voice for wildlife.

Planning permission may be given with environmental conditions attached, but these are frequently ignored and the authorities do not have the resources to battle with developers. Yet guidance that all new buildings should include swift bricks is starting to be adopted as official policy, often as a result of lobbying by local swift groups. Exeter's pioneering example has now been followed by Cornwall and Ribble district councils.

Stephen has worked with the British Standards Institute to develop regulations on nesting holes in buildings. This is an essential requirement if the construction industry is to be persuaded that swift bricks should be integrated into all new buildings. Conservation work depends on such apparently bureaucratic minutiae. It relies on pains-taking and targeted action to remove obstructions which threaten the future of our wildlife. It means forging friend-ships and partnerships with people to open their eyes to why it matters that our houses are not sealed off from nature, but can be part of it with living walls and green roofs – and holes for nesting birds.

Stephen works virtually full time in his voluntary role, talking with developers and architects, working with them to show how simple it is to make bird-friendly houses, giving advice on where best to install swift bricks. Ever pragmatic, he suggests these holes should be sited away from windows. 'If you have a family of chirruping sparrow nestlings right by your bedroom window, it can be quite annoying,' he says. And this is a man who loves sparrows more than swifts.

Swift bricks are a growing feature of new buildings in Exeter, including the city library, ten blocks of student flats and eleven schools. He has worked on schemes throughout the south-west, from Cornwall through Devon and Dorset to Wiltshire and Hampshire. Some big ventures are underway, with several large-scale housing developments planned which will include swift bricks built into homes surrounded by orchards, ponds and avenues of trees. Swindon, for example, is currently undergoing a significant expansion in job opportunities and is working hard to attract people to live there by building garden villages with green spaces integrated into their design. From Poole to Penzance, the nesting and roosting needs of swifts, sparrows, bats and butterflies are beginning to be taken into account, with around 10,000 special bricks already incorporated into the fabric of new housing.

The Duchy of Cornwall has wholeheartedly embraced the swift brick idea. Stephen worked closely with their estate surveyor, architects and a consortium of developers on a mixed-use development at Tregunnel Hill on the edge of Newquay, installing an average of one brick with integrated swift box per house. Following the completion of that project, the Duchy made the decision to integrate swift bricks into all future developments. This includes 4,000 new houses at Nansledan, an extension of the Newquay settlement.

At Stephen's request, it was agreed that the nest bricks should be monitored for ten years in order to establish which species used them and how effective they were.

Occupancy rates were fast, with a third of the boxes taken up in the first year, chiefly by sparrows but also a pair of swifts. The bricks were also incorporated into the brand-new Nansledan Primary School, and sparrows took occupancy before the doors opened and the first children skipped into the playground. Playing attraction calls is not always practical and, where there is an abundance of holes, it is not even necessary. Where sparrows go, swifts will follow, attracted by their unmissable cheeping.

At least 500 boxes have been installed in new houses built on Duchy of Cornwall land in Cornwall, Dorset and Oxfordshire. The Duchy has made a commitment to integrate an average of one swift box per home built on its land, which could lead to 5,000–8,000 swift boxes in place over the next thirty years. The personal passion for the environment of Charles, Prince of Wales and Duke of Cornwall, has driven the eco-friendly nature of these developments, but people involved in their construction are also proving influential in spreading the practice. The Duchy estate's former head surveyor is now advising a number of like-minded landowners on similar projects, so multiplying the numbers of swift-friendly buildings under construction. These include houses in a garden village on the edge of St Austell, Cornwall, and new buildings for the Dartmoor Whisky Distillery.

The proven value of swift bricks is now being recognised. Government planning guidelines issued in 2019 state that any new residential development should attain a 10 per cent net gain in biodiversity. One way of achieving this, the document states, is through the integration of a

universal bird brick for building-dependent species, providing nesting and roosting holes for a range of birds. This
would be based on the swift brick, which has proved to
be a magnet for birds that nest in buildings, such as house
sparrows – as well as swifts.

House sparrows are red listed, on account of their steep
decline over the last twenty-five years. Swifts, despite an
annual 5 per cent decline in recent years, are not – yet.
Providing nesting holes for both these species is vital but
the widely deployed sparrow terrace – a rectangular box
divided into several compartments – suits neither bird.
Sparrows, like swifts, nest colonially but prefer not to be
cheek by jowl.

Another crucial champion to the cause is the Royal
Institute of British Architects (RIBA), which advocates
integrating swift bricks at a ratio of one per new building.
One day, let us hope, it will be mainstream practice to
do this.

Stephen is passionate, too, about getting swift boxes
into church towers. With at least one in every parish in
the country, this is a simple way of ensuring the provision
of nesting places in both rural and urban areas. These
towers have the great advantage to swifts of height, and
it is easy to put boxes behind the louvres in belfries, while
excluding entry to the dreaded pigeon. With more than
30,000 churches in the UK, the prospects are potentially
exciting. Around the country, nest box projects in dozens
of church towers have already demonstrated the effectiveness of this approach, with colonies of twenty or thirty
pairs not uncommon.

While the details of holes in boxes and bricks absorb much of his time and attention, Stephen's vision is focused on the future. He is passionate, not just about providing nesting holes but also strengthening the natural environment to give us greater resilience to climate change, help reduce air pollution and bring health benefits for people. Climate change is destabilising our world, but there is much we can do to limit the effects. There are huge opportunities, not only in the way we construct new housing but also in the industrial sphere. Enormous sheds are being erected all over the country, warehouses and distribution centres which are not just a blot on the landscape but disruptive to the environment, causing loss of natural habitat and increased surface flooding. 'But they don't have to be like this!' Stephen exclaims. 'Include rainwater harvesting and living walls in the design, solar panels and green roofs, and the buildings start to earn their ecological place.' Such buildings could be gentle on the eye, friendly to wildlife and help clean the air.

Despite the huge challenges ahead, Stephen is steadfast in his optimism for the future. 'I am absolutely certain we will get there if enough people care and work together,' he says. The dream does not become reality all at once; change comes slowly, driven by determined individuals with a fine combination of passion and practicality.

18

MIDSUMMER

In the second half of June another big wave of swifts arrives. These are the immature birds, those that fledged last year or the year before. Curiously, a few of these non-breeding birds stay in Africa, but the great majority make the journey back and forth each year. Third-year birds appear to arrive back before the yearlings; they will need to defend their nests against the younger birds. Together, these non-breeding swifts are the most vocal; wild packs of screamers race over the rooftops in fine weather, circling around at top speed. It looks like sheer *joie de vivre*, the wild play of teenagers wanting to signal their presence, but is there also a purpose to it?

Close observation of these displays by swift-watchers in Switzerland and Germany has established that their principal function is social cohesion. The young birds are marking the territory of the nesting colony, learning its landmarks, such as trees, building and roads. Swifts maintain the same circuit boundaries year after year, generation after generation sticking to the same perimeters. Always, they revolve around the nesting colony. These immature birds are beating the bounds, just as people have done for centuries, around their parishes. While they continue to forage far and wide, these apparently boundless birds impose limits on themselves within the breeding area. It seems paradoxical, but for a short time this concept is a necessary part of their lives. Such boundaries help to reduce conflict when searching for nests, as the birds will look only within their own territory.

While swifts do not breed together in the density of, say, gannets or guillemots, they tend to nest close to each other, so a swift colony could be loosely defined as birds that nest within the same area. This might be in a single building or several nesting sites along a street or even half a street. Meanwhile, there may be another swift colony at the end of the road or a parallel street. Swifts interact with each other within their own groups. You can see this as you watch their screaming parties and chases, and when you listen to the calls between birds belonging to the same colony. By the second half of July, the boundaries start to erode and swifts from adjacent colonies will fly together.

Swifts gain a sense of belonging through these exhilarating flight displays. They may not yet have a nest but they are forming an attachment and staking out their place for the future. Breeders are often enticed out of their nests to join them, a social activity enjoyed by the whole colony, though these older birds are less frenetic and fly fewer circuits. After a few turns they return to their nests, keeping watch at the entrance to defend their holes against newcomers, often emitting a short, shrill scream when a pack of shrieking swifts goes by. Once the full clutch of eggs is laid, adult bird participation ceases; their time must be spent incubating the eggs, brooding young and finding food. The younger birds will soon start seeking and claiming their own place to nest.

It is not for nothing that we associate swifts with sunny summer days. These exhilarating flight displays take place only in fine weather. They occur throughout the breeding season, a thrilling sight to witness. Unless, that is, you happen to be a small brown bird, such as the dunnock I watched through the open door of my kitchen, quietly pecking about in the flowerpots. A searing screech rent the air as a party of low-flying swifts raced over; the bird froze for a split second, then fled.

The word 'screaming' is useful shorthand to describe their calls but an utterly inadequate description of their evocative cries, which vary greatly depending on the distance between bird and human listener. High in the sky they have a soft, silvery ring; milling overhead, the sound is like the shaking of tambourines. Lower down, their shrieks are piercing to our ears. Sometimes you hear a single call, sometimes the ear-splitting sound of a pack in full chase. Given how loud they can be and how suddenly they appear, it is odd that so few people seem to give them so much as a glance.

A friend, Jules Perry, related how he had been surprised by swifts in Istanbul:

Once, when exploring the area around the Galata Tower in Istanbul, I heard a sound like a beaten-up old car taking a corner very quickly, perhaps fleeing the police. Some minutes later I heard the sound again coming from a different direction. Again, no car. Then a second or so after that it was as if scores of invisible old bangers were all squealing round invisible corners. As soon as I looked up I immediately saw the answer – huge, pale brown Alpine swifts were appearing en masse. The sky seemed filled with these big swifts hurtling around, both near the ground and up around the top of the tower. I couldn't get enough of the sight. Yet all too soon they had gone. No one else seemed remotely interested, so I lost myself among the crowds and trudged off to get something to eat.

The calls swifts make will sound different to themselves than they do to us. Recordings of their calls slowed down reveal an unexpected diversity of single notes, rising and falling in pitch and with a continuous bubbling call. This sound has been described as like the clucking and crooning of hens on helium.

In the tower in Oxford, David Lack had noticed that swifts sing in duet – the *sweeree* call is the separate sounds of two birds. Settled in his Frankfurt attic, Erich Kaiser too listened attentively to swifts for many years. He noticed that the female sings the first, higher note *swee*, and the male the second, lower note *ree*. Thus, he suggested, the difference in the pitch of their calls makes it possible for the swifts to identify each other's sex. Their plumage – to human eyes at least – is identical. This discovery throws light on the behaviour of bangers* and fighters, as he wrote: 'Sometimes a single swift has occupied a box. Let us assume it is a male; then, if the banger, screaming into the entrance, is also a male, they both shout (to simplify matters) "Boy! Boy!" and if the stranger enters he will be attacked. If the banger is a female, however, she will shout "Girl! Girl!" while the owner will answer "Boy! Boy!" and this may be the beginning of a new pair formation.'

It is not always quite that straightforward, however, and fights between males and females have also been observed.

Using technology unavailable to Kaiser at the time of his study, fellow ornithologist Konrad Ansorge used spectrogram recordings of swift calls and revealed that the differences between male and female calls were not in pitch but in the rhythm of the trill that follows the scream. The male of the pair consistently showed a faster trill frequency than the female. There were indeed differences in pitch, but these could be attributed to the bird's degree of agitation.

Sometimes a scuffle breaks out before the birds have clearly identified their opponent's sex. Where this happens between opposite sexes, the different calls appear to calm them, ending the fight. Conversely, when a struggle breaks out between birds of the same sex, the calls provoke them further, increasing the tension. An example of the necessity of good communication was witnessed in a Welsh

* Bangers are immature birds searching for nesting holes.

swift box, where two young birds had fledged in the morning. In the evening, one of the parent birds arrived back in the box and settled on the empty nest. Its mate joined it shortly afterwards. This bird had clearly not accepted the departure of its young; it had brought a food ball in its throat pouch. The first bird became twitchy and eventually attacked its partner. A fight broke out. After some while this ended and they settled on the nest together. Could the fight have been caused by the second bird's inability to call and identify itself, due to the food ball in its throat?

In late summer, just before they leave on migration, clustered groups of screaming swifts are particularly frequent. Are these wild parties a sign of their restlessness to leave? Could they be a way of unifying the group before their 5,000-mile journey south? We do not have the answers, just theories. For us, though, 'the cut air falling in shrieks on our chimneys and roofs', as poet Anne

Stevenson describes it, is one of the most thrilling experiences of summer.

Second- and third-year birds may not be ready to breed
but they do not spend their summers in idleness. During
their brief sojourn, they search for a dark nesting hole
and find themselves a mate. Securing a nest site takes
time – black holes are not readily visible – but they have
an instinct for suitable places, usually attracted to them
by the sounds of their own kind. They will look for somewhere close by, but how to know whether or not the
hidden cavities are already occupied? A solitary young
swift, or sometimes a few of them following each other,
will glide silently around potential nest sites, then fly up
to them, brushing a wing or knocking against the outside.
They do this in a leisurely way, all through the summer,
exploring the possibilities.

Ornithologists have noticed subtle differences in the
behaviour of second- and third-year birds. By using
different colour rings on the legs of fledglings each year,
they have observed that second years simply caress the
entry to a swift nesting place with their wing, while third
years are more aggressive, knocking it with their feet or
chest.

In this way these 'bangers' find out whether or not the
holes they are inspecting are occupied: resident birds will
come to the hole entrance and scream at the stranger,
warning them off. If the banger does not heed the message
and alights at the entrance – behaviour more common in
third-year birds – a fight is almost certain. Like many
other species of bird, they are reluctant to engage in battle

and the resident swift will first try to scare the interloper off with a threatening display of screams and raised wings. If the prospecting bird refuses to budge, a fight starts. They rush towards each other and grip each other's legs with their sharp, powerful claws. Grappled together, these slow-motion fights often last for several hours, rarely less than twenty minutes.

Violent struggles are interspersed with pauses, during which the under-bird in the fight may appear exhausted, eyes closed and voicing a weak, piping call. Recovery can be very quick, though, and the bird with its back to the floor is often the victor, as this position enables it to shove the top bird towards the entrance with its feet – and push it out. Often, both birds fall out together, loosening their hold on each other as they plummet down, though sometimes they are found still locked together on the ground, where they are extremely vulnerable to cats and may face problems getting airborne again.

Surprisingly, they do not generally do each other much harm. While their claws are sharp, their legs are hard and scaly: chainmail armour. Wings become weapons with which to strike each other, and they jab one another with their beaks, though they may not inflict much damage beyond feather-pulling. They fight with their eyes closed, giving them some protection. Sometimes, though, it gets seriously bloody. Locked in combat, the chest is exposed and vulnerable to injury. Occasionally, the wounds are fatal.

The end of the fight is signalled when one of the birds gives a cry of submission, but it may be many minutes or even hours before the dominant bird releases the defeated one. This tenacity has been witnessed many times, notably on one occasion when an intruder was held dangling outside the nest for eight minutes, firmly in the grip of the resident bird's claws, its bloodied chest visible from the ground.

Swifts are also shown the way to potential sites by other hole-nesting birds, such as sparrows, cheeping noisily from the eaves. The swifts will often evict these small birds, piercing the eggs and carrying them off in their beaks to throw to the ground some distance away. When in competition with starlings, however, they generally come off worst. Starlings will fly onto the backs of swifts, knocking them to the ground. In Italy, however, swifts have been known to fling starlings' eggs from their nests.

Nesting holes are most often found in older buildings. Victorian and Edwardian housing provides the necessary cracks and crevices, but colonies of swifts are sometimes found in 1960s tower blocks and homes built in the last forty years. So long as the holes are there, they will find them eventually. Very occasionally, they have been known to occupy house martin nests.

While most swifts have a preference for nesting at height, giving their fledglings an extra split second or two in their first flight, others will breed in quite low places. The most extraordinary known instance of this is a pair that has successfully nested less than 42 inches from the ground, in the ruins of an ancient monastery in Northern Ireland. There are no absolute rules; every bird is an individual.

Once it has found a hole, the swift enters it tentatively, first poking its head through the entrance, then flying off. On its next approach the bird will enter a little further, up to its chest; after that it is bolder still and its whole body goes through the entrance. Finally, it will enter and look around. Once it has secured a nest it will defend it against all-comers, including birds of the opposite sex. For around five days the male will occupy the nest by himself. As he approaches his nesting territory, he makes a particular call, grating and sharp.

Courting swifts have their rituals. The male will approach the female with a quivering flight, with rapid, shallow wingbeats. Then he raises his wings in a V shape. These actions may be repeated many times, close to the chosen bird. If she is interested she will imitate these movements and then, with a shared energy and purpose, they will follow each other closely, cutting across one another, wings quivering. This stage of courtship is followed by a series of racing circuits in which they dash at more than sixty miles per hour above the colony.

Communication between swifts or other birds is impossible for us to understand definitively, but we can watch, listen and speculate. There is a particular call that the male makes, seemingly to advertise to passing females that he is a good catch. Two or three staccato bursts of sound tell her: 'I have a nest! Follow me!' Circling the colony, he makes a soft, low sound when passing the nesting cavity.

Next, he will demonstrate the flight approach several times. Flying at speed, they have to know their way precisely, and sometimes it takes a few attempts to reach the right position to enter the hole. She may not always follow him – not yet completely bonded perhaps, or unsure of her flight path. If this happens, the male will go out and find her – other males may already be trying to lure her away – and show her the way back to the nest site.

When the new pair finally arrive in the nest together they will preen each other for hours, making a distinctive soft sound, the 'I love you' call of the swift. After this, they spend their days scavenging the sky for nest material. This sounds difficult. Look up into the blue (or the grey), and you would imagine this an impossible task; there does not appear to be anything remotely useful floating around. Not to us, maybe. But the wind and even a gentle breeze blow all sorts of things into the air: feathers, thistledown, wisps of grass, tree flowers and sepals. Inevitably, in the twenty-first century, they also return with scraps of plastic wrapping. During World War II, David Lack discovered shreds of tinfoil dropped by the RAF to confuse enemy radar. This airborne flotsam and jetsam is then used to make the scrappy, shallow cup that is the swift's nest, glued together with saliva and shaped by the bird turning around inside, scrabbling with its feet.

Meanwhile, in nests close by, the breeding birds are combing the air for food for their fast-growing chicks. The next generation is getting livelier by the day. Nestlings are starting to exercise themselves, jumping up and down to strengthen their muscles, whirring their tiny wings rapidly – quite like a hummingbird. During their last few weeks before fledging they do press-ups. With wings outstretched and the tips touching the floor, they push themselves up until body and feet are raised above the surface, balancing on their wing-tips. Within a spacious nest box this is relatively easy, but many swifts nest in tiny, cramped holes with little room for their lengthening wings. By the time they fledge, their wingspan will be 15–16 inches.

19

EATING SWIFTS: ITALY'S TOWERS AND THE WAR ON PIGEONS

Common swifts have a vast range. From their wintering places in Africa, where they spend around eight months flying, feeding and rarely – if ever – touching the ground, they disperse thousands of miles east, west and north in order to raise their young. Across the world, they seek out crevices in churches, mosques, temples, castles, houses, libraries, restaurants, shops, factories and warehouses, integrating themselves into the fabric of our lives for around twelve weeks each year. Revered, ignored, unnoticed, studied or beloved, these birds are a living connection between people across the planet. The screeching of swift parties racing overhead and the softer calls of birds high in the air, their cries mellowed by distance, can be heard by billions of people globally, their piercing screams echoing over the chatter of hundreds of different human languages, a shared experience across nations and cultures, the soundtrack of our summers.

In each country, almost certainly, there will be a handful of people who become utterly absorbed by these birds, devoting time, energy and brainpower to watching and learning about them and doing practical things to keep swifts in our lives in the future. I decided to go and meet a few of them. This plan would dictate the shape of my holidays over several years, giving me a distinctive insight into the places I visited, their people and different cultural attitudes to wildlife. It is easy, when on vacation, to view our surroundings through rose-tinted spectacles, to be so overjoyed by sunlight, blue skies and beautiful old buildings that we suspend our critical faculties and assume we have arrived in a rural idyll. That is the idea of a holiday, after all. During my trips around Europe I would see some spectacular landscapes and bountiful wildlife, but with swifts forever in mind, my eyes would also be opened to environmental degradation and casual habitat destruction.

The skies of southern Europe ring with swift calls from late April to early August, for this is one of their main breeding strongholds. Vast numbers continue to delight us, the majority of which nest under the richly coloured terracotta tiles that have kept wind and rain out of people's homes since the Romans invented them. Increasingly, though, these are being blocked off with cement. There is no gain from this – in fact, householders are likely to find themselves with a damp problem and, ultimately, silent skies.

The renovation of houses, churches, towers and castles is taking place on a massive scale and swifts are being excluded from their nest sites. The urban war on pigeons, which also nest in buildings, is a major contributory factor. Pigeons are stupendously messy and their droppings can carry psittacosis bacteria, so excluding them is not unreasonable. Unfortunately, this usually results in all entry holes to buildings being netted or completely sealed, with the consequence that swifts and other small birds lose their nest sites, too. It does not have to be this way: a simple reduction in the size of the access hole is effective against pigeons and provides continued nesting opportunities for smaller creatures.

In Carmagnola, south of Turin, I meet Giovanni Buono. We walk past the shop that his parents once ran as a bar, the gold lettering on the window still bearing their name. Giovanni's bedroom was at the top of the building, directly opposite a church with swifts nesting in the stonework. As a child, he watched them, fascinated. A few doors up from his old home is a plaque to a renowned nineteenth-century intellectual, Giacinto Carena, a professor of physics at Turin University and a dedicated naturalist, who published, among other things, a study of leeches. He also campaigned strongly for protection of the Alpine ibex, which had declined calamitously, due to intense hunting pressures. Through a shared concern for the ibex, Carena forged a strong friendship with Franco Bonelli, a pioneering ornithologist, who in 1811 became the first professor of zoology at Turin University. His contribution to ornithology was recognised when both an eagle and a warbler were named after him. Bonelli is something of a hero for Giovanni and, with a twinkle in his eye, he imagines that as a contemporary of his great-great-grandmother, Bonelli might have encountered her while visiting his friend in Carmagnola. 'Maybe something might have happened between them, maybe I have his genes!'

Wherever his passion for birds came from, he has pursued it unswervingly, even learning English through reading bird guides. I find him in the town's Natural History Museum, of which he is the director. Immediately you cross the threshold, your eye is caught by swift silhouettes, stencilled onto the floor and stuck here and there on doors. Walk past glass cases of sparkling minerals, owls, butterflies and an armadillo, and you find yourself in a corridor with small, glass-walled offices. The last of these is Giovanni's – it has a silhouette of an Alpine swift, the biggest swift species found in Italy – stuck to the glass. 'Because I am the boss!' he laughs, office humour belying his modest personality.

We walk to the old Natural History Museum in Carmagnola, a tall building on the piazza in the centre, which exemplifies both the problems of overzealous reno-vation and practical ways of overcoming them. Some years ago, the museum was relocated and the building restored as an arts centre. Rows of holes are studded across the walls. These putlog holes were incorporated into cathedrals, churches and towers as they were built – poles were fitted into them to provide the horizontal beams in an early form of scaffolding often seen on the Continent. For centuries, such cavities have been occupied by swifts and other birds for nesting. The architect in charge of the renovation preserved the holes with meticulous care, respecting their historical significance. Unfortunately, he then allowed them to be netted to prevent pigeons getting in.

Dismayed by the loss of this thriving colony, Giovanni asked the owner of the neighbouring building if he could install swift boxes in a window at the top, inside an old pigeon loft. His grandfather had been a friend of the owner's family and agreement was readily given. The boxes were made and eight pairs of swifts moved in. Even better, Giovanni persuaded the owners of the arts centre to let him cut holes in the wire mesh fixed across the putlog holes: big enough for swifts but too small for pigeons. There are now more swifts nesting in the building than before the renovation. Even a plastic banner advertising the latest exhibition had not put the swifts off – 'they wait till the wind catches the banner and dive in behind it as it blows away from the wall!'

Giovanni is ever alert when the town's buildings are being restored and swifts are at risk. He speaks often to the planning authorities and they are always fulsome in their desire not to harm the birds. 'The problem is that unless you are constantly watching and checking, the swifts get forgotten – you turn your back and the ventilation holes are blocked off!' This is the same problem the world over and shows how the presence of such dedicated, unofficial swift wardens can make all the difference to the birds' future.

Giovanni's expertise is called upon by people in neighbouring towns and villages, who ask his advice on restoration projects. And word is spreading, with others becoming active in defence of swifts. Every town needs someone like Giovanni.

His particular interest, however, is the pallid swift, a species that first received scientific recognition in 1870, when an English naturalist, George Shelley, first separated it taxonomically from the common swift. As its name suggests, it is slightly paler than its better-known relative, with a whitish throat patch and chunkier wings. When Giovanni first began looking for them, they were much under-recorded and little studied. This gave him the challenge he was looking for and the opportunity to spend lots of time with the birds.

The behaviour of pallid swifts is quite different from common swifts. 'They not only have a Mediterranean distribution but also a Mediterranean character! These birds are quite unpredictable. Sometimes they fly low, sometimes they stay till November and sometimes they have two broods – but not always!' Giovanni describes them with the affection of someone who has watched them closely over years. 'Nor does the pallid swift demonstrate great fidelity – I have recorded one exceptionally fickle bird nesting with four different females over seven years!'

One of the best places to enjoy swifts in Piedmont is Saluzzo, a town with pre-Roman origins, perched on a hill with a dramatic backdrop of Alpine mountains. This is the place in which Chaucer's *Clerk's Tale* is set, the story of the cruel Marquis of Saluzzo, who tested his wife's love and patience by spiriting away their children and telling her they were dead. The town grew and thrived in the fourteenth and fifteenth centuries, leaving a rich architectural legacy in its cathedral, churches and elegantly arcaded palazzos. We stroll along the steep, cobbled streets, pallid swifts whizzing low overhead, flicking to the side down narrow passageways and up into the ventilation holes that pock the walls of the cathedral, sometimes resting for a split second vertically on the brickwork, where their lighter colouring is evident. They tend to choose holes below those of the common swift. An Alpine swift swerves past, its rapid staccato call ripping the air. Quite recently, they have begun to breed here and in other towns below their traditional mountain strongholds, adapting to man-made nesting places.

It is May and the swifts have not yet reached peak numbers. By mid-June, Giovanni tells me, the streets will ring with the calls of hundreds of swifts, the shrill cries of all three species ricocheting off the buildings. His research student, Erika Tomassetto, is with us. She has graduated with a degree in biology, but there are currently very few jobs or academic options that would make use of her qualifications, so she is studying for a tourism degree, hoping to combine her knowledge of birds with marketing skills to develop ecotourism projects. These spectacular displays of swifts, combined with guided tours around their breeding areas, would be a thrilling experience for tourists.

It would not be the first time Italians have put swifts to an economic purpose. Within a terrace of quite ordi-nary-looking houses in Carmagnola, a bizarre four-storey tower rises from one section of the roof, faced with scalloped tiles and numerous rows of bricks with holes to encourage swifts to nest. Nesting holes were added on three occasions to the Borgo Vecchio, as this nineteenth-century house is known, providing space for up to a thousand swifts. For hundreds of years it was common in Italy for people to build nesting holes into walls to attract swift colonies. A proportion of the swift chicks would then be harvested for food.

Giovanni gazes at the tower, commenting wistfully that the house would make perfect bed and breakfast accommodation for swift tourists. Unfortunately, the house is not for sale, but dreams like this may have their day.

A year or two later I am back in Italy, in the foothills of the Apennines in the region of Emilia-Romagna. It is June and people are selling cherries along the roadside, luscious fruits of several varieties: dark red, scarlet and yellow. The Panaro river valley is full of cherry trees; the blossom in April must be wondrous. Early fruit fetches high prices, but now it is late in the season and their abundance has brought such a drop in value that cherries are falling from the branches, left to the birds.

In the woods along a stream a nightingale sings short phrases of rapid liquid notes. Further on, I come across sulphur springs and a man filling numerous containers; these waters are renowned for their health-giving properties. It is a hot day, yet I am not tempted to quench my thirst here. It may be the elixir of eternal youth, but wrinkles seem preferable to drinking this rotten-egg-stinking *aqua minerale*.

Torre del Castellaro in Sassi di
Roccamalatina national park in Italy

Out of the woods and into the meadows, massive sandstone outcrops appear, jagged-toothed pinnacles of rock,
rising dramatically above the trees. This is Sassi di
Roccamalatina, a national park, designated on account of
its magnificent rocks, rare plants and an array of wild
creatures including porcupines and wolves.

I wait outside a café, with a panoramic view across the
valley. Somewhat incongruously, it is called Il Faro – the
Lighthouse – despite being more than a hundred miles
from the sea. Soon, Mauro Ferri arrives, a retired vet with
a passion for and great knowledge of swifts. He is going
to show me round the Torre del Castellaro, an ancient

fortified tower, valued in changing ways throughout its existence. He gets out of his car brandishing an enormous iron key and we saunter up the quiet road towards the tall, square, whitewashed building.

Built in the thirteenth century for defensive purposes, you can still see the faint outline in the stonework where the original door was positioned, halfway up. Soldiers would climb a rope ladder to get inside, then haul it up behind them at night. Small apertures enabled them to keep a lookout for invaders, and fire arrows or throw spears when threatened.

When such fortifications became redundant they were often converted into swift towers, where colonies of the birds would be enticed to breed and a proportion of the chicks harvested as food. This was no mere passing fashion. Mauro's research has revealed that at least 10 per cent of the fortified farms and other historic buildings in the region had swift towers. Rows of holes in the tops of buildings can be seen in palaces, farmhouses, barns and belfries, and sometimes incorporated into dovecotes and belvederes. A fresco in the Church of the Sorrows in Lombardy, dated to 1470, illustrates a swift tower remarkably like the one we are looking at.

The practice continued over centuries in northern Italy, with large estate owners employing a swift keeper to collect their ration of young nestlings and clean out the nest holes after the swifts' departure. Each pair had to be left with one chick to raise, to ensure the future of the colony. Meanwhile, the keepers were allowed to eat the *pippione*, the chicks of doves and pigeons.

Mauro points out the three rows of holes near the top of the tower, which were added hundreds of years apart, complementing the existing arrow-firing holes. Swifts would make their nests in spaces within the walls, which were shut off from the inside with wooden plugs. Chicks would be taken at around twenty days old, to be turned into puddings, stored in terracotta pots and eaten during Christmastide, or given away to friends and honoured acquaintances. Described as little balls of butter, on account of their great fat stores, they were considered a great delicacy.*

Swift towers gradually fell into disuse in the first half of the twentieth century as people's attitudes changed – perhaps they discovered tastier things to eat. Legal prohibition came much later. For the last thirty years, Mauro has explored the history of these buildings, mapping their whereabouts, exploring archives and talking to the

---

* Lazzaro Spellanzani, an eighteenth-century professor at the University of Pavia, made numerous observations of swift behaviour within swift towers and even carried out an early form of bird ringing. He tied red thread around the legs of a pair of swifts, one of which returned the following year, demonstrating the fidelity of the bird to its nest site. Another time, he calculated that the swift's vision is accurate from 314 feet, by watching a flock of the birds feeding on flying ants and measuring the distance between a fixed point (a tree) around which they rushed, to the anthill from which the insects emerged. Spellanzani unequivocally dismissed the theory that swifts might hibernate.

A man of great curiosity and learning, he was widely renowned during his lifetime, gaining international recognition in 1768, when he was appointed a fellow of the Royal Society. An excellent teacher, he had a following of 500 students in one year. His studies ranged from experiments with microbes, observations on the navigation of bats and the process of animal digestion. He was the first person to perform in vitro fertilisation, which he carried out with frogs. He also unravelled the science behind stone-skipping.

last of the swift keepers. In 1986 he met an old man who had managed the colony at the Torre del Castellaro until his retirement many years previously. He talked of how he had been allowed to keep pigeons alongside the swifts, and how he was permitted to take these for himself. By then the grandeur of the tower was just a memory; seriously dilapidated, its roof was on the verge of collapse. Mauro, though, was inspired. That very day he rang the tower's owner and asked to rent the building.

It was a huge project. He repaired the roof and nesting places, reopening the holes, many of which had been blocked. A handful of swifts were still present but most of the remaining cavities were occupied by starlings, which had filled the nesting spaces with stacks of hay and leaves. When local people heard that Mauro would be restoring the tower for swifts, they were delighted. Starlings are regarded with suspicion and hostility in this area on account of their fondness for cherries – especially the highly-prized early ones.

Dozens of swifts reoccupied the tower, thanks to Mauro's efforts, along with lesser horseshoe bats and geckos. The building is now in the hands of the Sassi di Roccamalatina park, but Mauro continues to monitor and ring its swifts each year.

The temperatures are soaring outside but within the tower it is refreshingly cool. Wooden ladders are placed between the floors, with a rope to hold on to. 'Keep to the sides of the struts, don't step in the middle!' advises Mauro. A little warily, I climb up. On one wall there are rows of little Perspex doors covered in black paper. Swifts like darkness. Many of the chicks are close to fledging, a month ahead of those in the UK.

From time to time the tower is alive with the soft clamour of chicks as an adult returns, its throat bulging with food. These excited sounds have very likely been heard in this building for 700 summers. Mauro is planning to ring the young ones the next day, with the help of two other trained ringers. The job will take all afternoon; the colony has grown over the two decades since Mauro took it on, with a total of ninety pairs this year.

He opens one of the doors, lifts out a nestling and puts it in my hands. Warmth flows from its body into my palms, reminding me of the swift I had looked after a couple of years previously. This one has a restless energy that I never saw in my swiftling. In just a few days it will leave its dark hole and take to the air. In a fortnight, with luck and a following wind, it will be feeding over a steamy African forest. This young nestling is bigger than its parents – close to 50 grams, Mauro reckons. A hundred years ago, such a plump chick would almost certainly have been selected for a pudding. Not this one, but it will need to lose a few grams before it can fly. As Mauro takes it from me to return it to the nest, I notice one of the blood-sucking *Crataerina* has crawled from the bird onto my arm. It is the size of a large woodlouse. For a swift nestling this is a giant monster, yet they tolerate them, generally unharmed by their blood-sucking habits. The monster creature crawling on my skin is an initiation into the swift's world. Mauro, somewhat less spellbound, flicks it to the ground and stamps on it hard.

Every year, when all the swifts have gone, he will clean out the holes to reduce *Crataerina* infestation but replace the nests. They will be used again, with a few feathers added, next year. Until now, Mauro has discouraged species other than swifts from nesting in the tower, but the colony is now so strong he has decided it is time to take a more relaxed approach. Next summer, for the first time during his caretaking, he will not prevent great tits and other birds nesting in the top row of holes.

Managing a place specifically for one particular species is not exceptional. Conservation habitat plans inevitably favour one kind of vegetation over another: if you want to maintain wetlands you must remove bushes that encroach; to encourage summer flowering herbs you remove grazing animals from fields later than if spring species were a greater priority. Human intervention makes subtle changes such as these in the most benign environmental schemes. Nature, in our human-dominated world, has to be deliberately cultivated.

Leaving Torre del Castellaro, Mauro embarks on a tour of the local area, pointing out many other swift towers, most of them incorporated into farmhouses, others in stand-alone buildings. The holes have been sealed off in many of them and he is deeply concerned about the loss of these ancient, purpose-built nesting places. Hundreds of generations of swifts have been raised in them, keeping healthy colonies in countryside, villages and cities. Here and there he has found owners prepared to reopen them when they learn the history of these holes and how much swifts need them.

A little truck with a shiny steel tanker on the back races past us, just before a bend. Mauro shakes his head. 'It's carrying milk and will be on its way to the Parmesan factory – it has to arrive fresh from the cow, they cannot use milk that has been refrigerated. You see these tankers speeding along all the time.'

He has been involved in the swift-friendly restoration of many other buildings, including the belfry of Modena Cathedral, the 300-foot-high Ghirlandina tower. The authorities had planned to seal all 200 scaffolding holes in this medieval building to exclude pigeons, but Mauro and his veterinary colleagues suggested that they simply reduce the size of holes, keeping them accessible for swifts and other birds and bats – but not pigeons. This would, they advised, be a fabulous way of celebrating the European year of biodiversity in 2010. Thankfully, their idea was accepted and the swifts were saved.

Entire colonies of swifts, bats and other creatures can be entombed when buildings are renovated during the nesting season. The consequences are almost as serious when work is carried out during their absence and the site-faithful birds return to find their nest holes blocked. Mauro is campaigning for procedures to be adopted, applicable to both old and modern buildings, to stop this happening.

Back at the Lighthouse café we sit outside for a while, sipping ice-cold beer. Day-flying moths flicker around us, one landing on the rim of my glass. Black with half a dozen or so bold white splodges on each wing and with white-tipped black antennae, it is a handsome creature.

'We call them little friars,' says Mauro. The keening wail of a peregrine falcon sounds from below, one of a pair breeding on the rocky pinnacle, close to where a hermit once lived.

Mauro tells me more about bird harvesting, gleefully pointing out that the English, too, had indulged in this 'sinful behaviour'. Shards of red clay bottles, believed to have been sparrow nesting pots, have been discovered during archaeological digs in London, Kent, Hertfordshire, Essex, Buckinghamshire and Cambridgeshire. These pots have a so-called 'robbery hole', enabling a human hand to reach inside and take the eggs or nestlings. Young sparrows, like swift chicks, were cooked in pies and tarts. The superfood of their time, it was said that they made the eater more sharp-witted.

Since the Middle Ages, terracotta bird pots were used across northern Europe, deployed in the Netherlands to attract both sparrows and starlings. The painting of *St Christopher Carrying the Christ Child* by Hieronymus Bosch, dated to 1496, is full of curious landscape details. As the old man toils up the hill away from the river he has just crossed with a child grown as heavy as the whole world, he is shown among cultivated fields, with a peasant woman, some domestic fowl and a small white dog in the foreground. A beehive is fixed to the top of a dead-looking tree and below this a large sparrow pot hangs from a branch. In a somewhat whimsical touch, a tiny ladder to the hole is depicted, presumably for the sparrows to climb. These pots (most of them probably without ladders) were exported all over Western Europe until the twentieth

*Saint Christopher Carrying the Christ
Child* by Hieronymus Bosch, 1496.

century. Yet while sparrows were a useful addition to the diet of poorer people and a gourmet delicacy for the rich, they also earned a reputation as vermin. From 1532 a reward was paid for catching sparrows, which gave people a new incentive to breed them. They could also be honestly sold to falconers to feed their hawks.

Sparrows were harvested in Italy, too, with colonies encouraged by incorporating holes into buildings, much as they were for swifts. For sparrows, the holes were closer together, a whole grid of nesting cells integrated into a wall. A few buildings had provision for both swifts and sparrows. No doubt sparrows attempted to nest in the swift holes, though their attempts may been foiled if the owner was set on making swift puddings.

Terracotta bird pots have made a comeback in recent years, sold as nesting places for garden birds such as blue tits and sparrows. Today, the broods are safe from theft by human hands, but face all the other hazards of the twenty-first century.

Looking across the wooded landscape below, Mauro talks about how the area has changed over the last fifty years. Many smallholdings have been abandoned as young people migrate to towns and cities and the older generation dies. Not so long ago these hills were much more open, with cultivated fields. Now there are chestnut trees, downy oak and hornbeam – a great source of insects, says Mauro. Traditional farming would almost certainly have encouraged a greater range of invertebrates, but if the land had been taken into intensive agriculture this plentiful source of food would have gone.

Staying in an apartment in an old farmhouse, I have seen the richness of the old meadows. The grass crackles with insects: butterflies, moths, grasshoppers and many kinds of tiny beetles. Cranking up a window blind on the first morning, I saw a hoopoe perched on the telephone wire, snapping its long, down-curved beak. These grasslands provide bountiful pickings for them. Another morning brought a glimpse of a leveret, its black-tipped ears swivelling above the tops of the grasses, listening.

The conversation shifts to Italy's obsession with hunting.

Its popularity is declining, but from a very high peak. There are still 800,000 registered hunters, down from 2 million a few decades ago. Each year they shoot or trap more than 5 million wild birds, including rare species, many of them illegally. The trapping of small birds for the restaurant trade is also common, and many of these are smuggled in from other countries, notably Serbia. While there is growing revulsion for the practice among many Italians, killing and catching wild birds remains deeply embedded in traditional culture and attempts to curb it are regularly thwarted or undermined by government. Centuries of persecution have made songbirds extremely wary of people. The robin that will come and eat cheese from your hand in England or Sweden is a skulking creature in Italy.

Grim though the spring and autumn carnage of migrants is, the biggest problems for birds and other wildlife are intensive agriculture and habitat destruction. The latest manifestation of humanity's boundless facility for finding new ways to create environmental havoc can be seen in the river plains of Emilia-Romagna, where willow, oak and poplar are being felled along river banks and harvested for the biofuel market. This is not skilfully-carried-out coppicing but mechanical extraction, which wrecks the visual landscape, knocks out whole ecosystems of plants and creatures, and leaves the rivers vulnerable to flooding, their banks weakened without the holding power of tree roots.

But 1,000 feet up in the Apennines on a glorious evening, all seems good with the world. Mauro talks of the international swift festivals he started a few years ago. Sharing the joy of swifts alerts people to their habitat needs and makes them want to do something about them. Throughout Italy, in villages, towns and cities during June and July, swift walks are held to watch the birds – Alpine, common and pallid – race overhead in the evenings. The idea has caught on and swift festivals are now held in other countries, including Switzerland, Israel and the UK, raising awareness of the thrilling phenomenon that occurs each summer and showing how people can do their bit to ensure it continues long into the future. One man, inspired by what he saw and heard at an event, returned home and made seventy swift boxes. He then persuaded local firemen to put them up in nearby villages.

Into the golden light a rush of screaming swifts appears, hurtling around the tower and over the woods below. I return to my *agriturismo* lodgings and, as the sky darkens, take a stroll through a grove of trees. Little flecks of light attract my attention. Fireflies! Weaving their way between the trees, they move slowly, two or three feet above the ground, their moon-bright luminescence flashing on and off, a signal to potential mates.

20

PIONEERING BIRDS AND DETERMINED PEOPLE

Southern Spain offers something unique in Europe – the possibility of seeing no fewer than five species of swift: common, pallid, Alpine, white-rumped and little. While I had already encountered pallid and Alpine in other parts of Europe, I was looking forward to seeing them again; it takes many encounters to get to know your bird. Alpine swifts breed in a broad sweep across southern Europe, eastwards through Turkey, Afghanistan and on to the western side of Bhutan, while pallid have a more restricted and scattered range around the Mediterranean, Portugal and the Persian Gulf.

White-rumped and little swift are relative newcomers in Europe. Both have their main centres of population in sub-Saharan Africa, while little swift is also strongly established in south-east Asia. Little swift colonised Morocco in the 1920s, spreading north to the Tangier Peninsula on the Strait of Gibraltar in the 1950s. Its arrival in Spain was strongly anticipated; the distance between the two countries at the narrowest point is just nine miles. However, there were no such expectations for the white-rumped swift; no one knew that they too had extended their range to North Africa. So when a pair of swifts with white rumps appeared in Cadiz in the early 1960s, bird-watchers jumped to the conclusion that these birds must be the long-anticipated little swift. It was only when a photograph of the birds appeared in ornithological magazines that the error was put right: these swifts were *Apus caffer*, the white-rumped swift. Little swifts, *Apus affinis*, finally made the leap to Spain in 1992 and bred in 2000. Numbers of both species are still low but rising gradually.

So it was that I arrived in Seville at the end of May to enjoy ten days in the company of swifts. I had booked a room in the medieval heart of the city, a maze of narrow streets and plazas full of orange trees. Thirty-seven steps up and the door opens into a small, clean bedroom with old-fashioned fittings, painted the blue of a chaffinch's head. Wooden glazed doors lead to a small terrace and then, joy of joys, a spiral staircase to the roof. Dozens of swifts are racing their circuits, gliding above the pantiled roofs towards the Giralda, the cathedral bell tower, and swooping up to the eaves of the old hospital of venerable priests, a seventeenth-century building that now houses the Velázquez centre; the painter was born in Seville.

Rich in colour, the old terracotta tiles are mottled with a patina of lichens. Blue-glazed ridge tiles add a nice detail. Who would see these, apart from the few rooftop visitors? Ominously, I notice that many of the outermost pantiles are blocked with concrete, sealing off the entrances to holes where swifts and sparrows nest. This obstructive practice is becoming increasingly common; people really do like to keep nature at bay.

Fortunately, there are two women in Seville speaking up for swifts. A text from Elena Portillo tells me she will be a few minutes late for our meeting. She is at the police station, reporting the destruction of a large colony of nesting pallid swifts in Toledo. She arrives a few minutes later, with her mother, Esperanza. 'Nest blocking is a big problem in Spain and the police don't care, even though it is illegal!' With a network of swift-watchers around the country connected via social media, she is encouraging people to gather evidence of the birds' nest sites, without which it is impossible to defend them when they are threatened.

Nor do many residents or local politicians take seriously the problems encountered by swifts, she tells me. Elena, however, is on a mission to change attitudes. Young, beautiful and articulate, she is an excellent champion for the birds. Her interest was nurtured by her mother, who is also passionate about wildlife. Both are ardent campaigners for swifts and other birds in the city.

Aged eleven, Elena found an injured house martin and tried to save it. 'Back then, I was ignorant and fed it the wrong food and the bird died.' The experience sparked

Elena and Esperanza Portillo.

her interest, though, and she has continued to rescue young birds, rearing them successfully with the benefit of experience. A few days earlier she had come across some children throwing stones at house martins' nests. She admonished the culprits and took in the birds; now they are thriving. And the rescued martins have become their own advocates, as Elena posts photos of their progress on social media, raising awareness of wild birds and the problems they face.

She works as a science communicator at the university, and in her spare time visits schools to spread the word about swifts. 'Young people are very receptive and want to help but old people don't want to know. They like everything to be clean and tidy and don't want to have birds in the cities!' She also talks to architects and the police, attending conferences and speaking to them directly.

A perception that wildlife can look after itself is prevalent. With such an abundance of swifts in the city, it seems most people simply take these birds for granted. The Spanish bird protection organisation, SEO Birdlife, has just 11,000 members. Compare that with the UK's RSPB, with a membership of over a million. There is no natural history museum in Seville, but Elena has ambitions for an environmental education centre, where people can come and find out about the rich natural environment of the city.

During the week leading up to Easter, Holy Week processions are held every day in Seville. Organised by religious brotherhoods, ancient wooden sculptures are paraded on floats, presenting scenes from the Passion. The festival is of great significance, both spiritually and culturally. Crowds of people take to the streets. 'Everyone falls silent and all you hear is the sound of swifts,' says Elena. Swift calls have become inextricably associated with the processions, she tells me, yet people do not think twice about blocking off their holes with concrete, often trapping adult birds incubating eggs or nestlings. 'Only when we have no swifts during Holy Week will they realise what we have done,' she fears.

At Elena's suggestion, we stroll off to find Las Setas de la Encarnación. This surreal flight of fancy is a wondrous construction in birch wood composed of flowing, mushroom-shaped structures over 80 feet high, extending across nearly 500 feet of open space. Below ground is an archaeological museum housing Roman and Moorish artefacts found on site, while the plaza is the venue of a long-established market and a cool and airy public space.

A group of teenagers are somersaulting and whirling on the ground, practising their break-dancing moves. Further on, a scattering of young lads are playing football, while babies in buggies are wheeled along by their parents. Above them, dozens of pallid swifts circle under the sweeping curves of the mushrooms, their joyous antics another layer of activity in this inspirational place, which has provided them with an array of new nesting opportunities. The roof is composed of hundreds of box-like structures, with many little gaps between the timbers, perfect accommodation for swifts.

Between two forks of the River Guadalquivir, south of the city centre, is a sizeable island that was developed for residential accommodation over the second half of the twentieth century. This is the *barrio* of Los Remedios, with a long, straight avenue of multi-storey apartments and shops, built in different phases and styles. Intensely urban, it does not immediately appear to be an ideal place to find swifts, although the nearby river provides good feeding opportunities. Yet nowhere have I seen a higher density of these birds; the air is thick with them, their shrill calls rising above the engines of cars. Only around the newest blocks are they absent. Nor is it only swifts that find homes in this man-made canyon. Common kestrels are often discovered in window boxes and flower pots on the highest balconies, safe places to raise their young.

Darkness has fallen by the time I return. Around the floodlit cathedral swifts can be seen flying around the bell tower, hunting moths in the company of bats and lesser kestrels, for whom this building has long been a stronghold. I stand awhile, enjoying the spectacle until an American tourist asks me 'Is someone trying to jump?' When I answer that I'm watching swifts, he looks blank and walks away.

The next morning I watch pallid swifts on the east side of this vast building, darting back and forth, close to an intricately carved archway, their cries interweaving with the clanging of the bell above. One of them disappears behind an angel into an invisible crevice – just the thing for a nesting swift.

A hundred miles or so north of Seville is Mérida, principal city of the ancient Roman province of Lusitania, which stretched across much of what is now Portugal and western Spain. Today it is the municipal hub of the autonomous community of Extremadura. An abundance of Roman architecture has survived: a magnificent theatre and amphitheatre; statues (by no means all of them headless); porticos; pillars of a ruined aqueduct – ideal nesting platforms for white storks; the world's longest surviving Roman bridge with sixty-two spans across the River Guadiana; and the National Museum of Roman Art. Ringed by blue and white road sign arrows, a broken section of a column and its decorated capital occupy a mini-roundabout, traffic circling in a bizarre juxtaposition of ages. Faces of the rich and powerful, ingeniously carved some 2,000 years ago, look down from a temple pediment to the people below, their ancient authority still palpable.

With its subterranean vaults and series of Romanesque brickwork arches in the main exhibition space, the museum offers airy respite from the afternoon heat. I look in vain for swifts in the carvings. Some crane-like birds are carved into the frieze of a stone coffin and three doves flap their wings in an embossed tree, but I find few other natural details apart from decorative foliage. The Romans, like us, were more interested in their own species. I am curious to see an earthenware whistle, shaped like a hen, highlighted in the museum leaflet. I remember buying such a thing for my own children, though made of plastic. Pour in a little water, blow, and it makes a fluty warbling sound. It is missing from its glass case and the attendant is unaware of its existence. I hope it has not crumbled to dust.

In a café outside I meet Jesús Solana, a vet with the regional government. He has a lifelong interest in wildlife but fell under the spell of swifts when he learned about their awesome lifecycle. He tells me a story about pallid swifts. Ten years previously he had seen a group of nine flying over the village of Alange, seventeen miles from Mérida, on 5 January. The next day they were there again. The following winter he returned and once again saw swifts – fifteen of them. This was unheard of. Pallid swifts often raise two broods, sometimes staying in their breeding grounds until the third week of November, but never, it was held, throughout the winter. I remember Giovanni Buono, my friend from Carmagnola, telling me that unpredictability is in the nature of pallid swifts. Jesús was determined to gather evidence of the phenomenon, so every winter since he has made frequent visits to Alange, recording between two and eight swifts every year but one. The birds roost in holes in the church, entering in pairs. 'It is strange to see swifts entering the church while the Christmas decorations are there!'

I have told Jesús I am hoping to see white-rumped swifts. 'It isn't easy,' he tells me. Several other people have already warned me of this. The European breeding population is tiny, with just a few hundred pairs in Spain and a handful, possibly as few as ten pairs, in Portugal.

'Let's go and look – you might see some at Alange!' Jesús says, still enthusiastic after a long day's work. We walk up a rocky hill to the ruins of a ninth-century castle. Along the path the dry husks and stalks of oat and other grasses shimmer with light from the setting sun. Spanish broom and honeysuckle scent the air and a lark – 'Thekla!' says Jesús – sings overhead. Wild artichokes are in flower: 'The petals are used in cheese-making,' he tells me. They contain an enzyme with milk-clotting properties, so have been harvested by cheesemakers for centuries.

Below the castle keep – all that survives of this Moorish fortress – the ground opens out, with views down to a vast reservoir with hilltop islands, the water golden in the dusk light. We watch silently for a while, looking out for a swift that is slighter than our common swift, with long wings, a forked tail which will be closed in flight to a needle-thin spike, a distinctive white chin and a rough crescent of white on its rump. House martins, red-rumped swallow and a couple of common swifts are snapping up insects, but *Apus caffer*, the white-rumped swift, is nowhere to be seen. It was never going to be that easy.

We retrace our steps down the hill and head for the Church of Our Lady of Miracles. The air is thick with swifts; hundreds of them nest in crevices in its walls. *Apus pallidus* is highly gregarious, nesting in tight colonies. The difference between their calls and those of the common swift is subtle, but there is a characteristic down note, which I am starting to catch. By the dam, Alpine swifts are whirling around, white bellies visible as they turn, uttering their loud, unmistakable metallic trills. More than 1,000 of them have been counted here in a single evening.

There is indeed something quite miraculous about Alange: it is a magnet for swifts. With its phenomenal wintering pallids and breeding populations of Alpine, common, pallid and a few elusive white-rumped, this village is exceptional. It is also the only place so far in Extremadura where little swift has been recorded – just once, in 2012. When Jesús decided to organise a summer swift festival, Alange was the obvious place to hold it. With walks and seminars, photography workshops and activities for children, it has grown in popularity each year. Residents are realising the value of the swifts, recognising that these birds are an asset to their village. While Alange has long attracted tourists to its Roman baths, fed by thermal springs, people are now also drawn here by its spectacular swifts.

Raising awareness of swifts and their problems is as crucial here as in Seville. The threats are seldom obvious but Jesús tells me that colonies are highly vulnerable to destruction. He has set up a survey group to locate swift colonies in castles, churches and public buildings in order to try and ensure the survival of nesting holes if and when renovation is undertaken. This has included working with a school, where a large colony of swifts nests in the gaps at the top of exterior window blinds – commonly used niches in southern Europe. By good fortune, the head teacher is a passionate environmentalist and supports his plan to retain the swifts' nesting places by restoring instead of replacing the wooden boxes that house the blinds.

Not everyone is so receptive. A large swift colony in the city walls of Cáceres, a beautiful town some forty miles away, was being immured that very summer. Restoration work was underway in full knowledge of the birds' presence, their nest holes filled in.

Preserving the holes is perfectly compatible with the restoration of historic buildings; it simply requires someone to care. Jesús is striving to protect hundreds of Alpine swifts that nest under a nearby motorway bridge, which is currently undergoing modernisation. Getting the construction company to put up an adequate number of swift boxes – and install them in the right place – is proving difficult. 'Unless someone is watching and making a nuisance of themselves, nothing happens.'

He is deeply anxious too about the attitudes of local politicians towards the environment. Each of them seems to try and outdo the others in their antagonism towards environmental protections, which they see as obstacles to progress. He points towards a distant mountain, partially obscured by a thick plume of smoke. Controversially, planning permission was granted for a huge biomass factory and the filters that could have been installed were declared unnecessary. He shakes with fury while talking about it.

Extremadura is one of the most biodiverse areas in Europe.

With mountains, woods, rivers and vast areas of steppe, it
contains a mosaic of habitats and thus a great diversity of
wildlife. That this jewel of nature has survived into the
twenty-first century is a marvel. A string of international
environmental designations demonstrates its value; they
should not be regarded as a hindrance to prosperity. Amid
the dire reports of global species decline and extinction,
Extremadura can hold its head high. It should be – and no
doubt is for many people – a matter of great pride that the
region has escaped the ravages of industrialisation and urbanisation, something to be cherished and upheld. Smoking
chimneys belong in the past, not the future.

A birding friend who has made a home in Extremadura
suggests a visit to Monfragüe National Park, 100 miles
north of Mérida. 'There will definitely be a few white-
rumped there!' he tells me. With its forested hills and
mountains, deep river gorges and craggy cliffs, Monfragüe
is a spectacular landscape, rich in wildlife. To the north
and south are vast areas of *dehesa*, open groves of evergreen holm and cork oaks, with the land between
rotationally grazed, cultivated and rested. This ancient
system of agriculture is now providing ideas for modern
wood-pasture systems in the UK – sylvo-agriculture, as it
is called by today's practitioners.

In birding circles Monfragüe is internationally renowned
for its raptors. About half of all black vultures nesting in
Europe are found there, along with hundreds of griffon
vultures, Egyptian vultures, Spanish imperial eagles and
Bonelli's eagles. Whether I catch up with my swift or not,

I am certain of seeing some fabulous birds. Many of them breed on Peña Falcon, the dramatic rock face above the River Tagus, its jagged pinnacles formed of massive slabs, upended at some distant point in geological time. Hunched together on its narrow ledges, young griffon vultures utter short, rasping shrieks. I see the shaggy brown feathers of their wings and bodies and the pale, almost fur-like plumage on their necks and shanks. Just beyond them is an even bigger bird, the cinereous vulture, also known as the black or monk vulture. Extinct across much of its historic range in both Europe and north-west Africa due to habitat loss, poisoning and shooting, there are now more than 300 pairs in Monfragüe.

With just a pair of binoculars, though, I am ill-equipped to see the details of their faces. Birders cluster in the lay-by with tripods, telescopes and enormous telephoto lens cameras focused on the rock face across the river. One of them invites me to look through his scope. The vultures' beaks, chunky and fearsomely sharp, come into vivid focus, perfect for tearing into the carcasses of the beasts on which they scavenge.

It is not simply a matter of geography that birds of prey thrive here. Probably the main factor is transhumance, a traditional farming practice that mimics migration. Twice a year, livestock is herded over long distances along ancient droving routes, from winter to summer pastures and back again. The sheep and cattle involved are tough breeds, well capable of the long journeys, but inevitably some will die through age or illness. When this happens the carrion is swiftly eaten by vultures. Inadvertently, these peregrinations perform another service for wildlife as the animals spread the seeds of wild flowers and grasses that stick to their hooves. The pastures along drove roads tend to be particularly rich in species.

You might think that a place as outstandingly biodiverse as Monfragüe would be inviolable, that destructive commercial operations would never be countenanced. Such foolish trust would of course be misplaced. Despite recognition of its rare ornithological interest both in government circles and among conservationists, plans were drawn up in the early 1970s for a paper mill upstream. Eager to cash in on the opportunities this presented, landowners drew up plans for a major clearance of the native vegetation, which would be replanted with eucalyptus and pine.

Bizarrely, it was the Department for Nature Conservation that drove this destructive idea forwards. Staffed by individuals with strong connections to the forestry industry, wild nature stood little chance. Soon, access tracks were carved out of the forest and the felling and replanting began, aided by grant funding from the Ministry of Agriculture. Catastrophic destruction was averted by a huge public campaign, with international backing, led by conservationist Jesús Garzón, renowned today for his passionate dedication to the natural environment and a dynamic advocate of traditional pastoral systems. Monfragüe was awarded natural park status in 1979, became a UNESCO Biosphere reserve in 2005 and a national park in 2007.

Early the next morning I take a steep little path to another ruined hilltop castle – there are hundreds of them in

Extremadura. A cuckoo is calling across the valley but there is no sign yet of the circling vultures that were so visible yesterday. The sun will soon heat the earth, columns of warm air will rise and raptors slowly spiral, scanning the ground for food. A blue rock thrush is perched in a tree just below the old stone walls. It flies, and sunlight turns its slate-dark plumage a shining blue.

'Good morning!' a young man with a German accent greets me. He is on holiday from Saxony with his parents, he tells me, and when he hears I am hoping to see white-rumped swift, he says that his father, who comes from Wolverhampton, wants to see one too. Circling the tower ruins for the view likeliest to reveal my bird, I come across a man with Zeiss binoculars – this must be him. Expecting to hear familiar Black Country speech, I am astonished when he answers my English greeting in as strong a German accent as his son's. After thirty years living away, he is speaking his mother tongue as a stranger.

His bird identification skills are formidable and I feel my chances of seeing the white-rumped swift rise. We look across the valley towards Peña Falcon, scanning the rocks, grassy patches, trees and the forking river below. There are lots of white-rumped birds around, all of them house martins. Watching them in the distance is like following specks of dust in a sunbeam, light glinting off their pale patches as they twist and turn.

White-rumped swift arrived in Spain in the slipstream of another newcomer, the red-rumped swallow. Throughout its range in Africa and now south-west Europe, white-rumped swifts make use of other birds' nests, sometimes

entering them when the original nest-owner has already
departed, sometimes evicting them. Red-rumped swallows
first appeared in southern Europe in the 1920s, expanding
rapidly in the 1960s and 70s. This swallow builds a mud
cup nest with an access tunnel, tight to the ceiling of caves,
rock clefts and open sheds. The tunnel varies greatly in
shape and length and tends to be shorter when built in a
sunny position, to prevent the nest getting too hot, and
longer and even curved on shadier sites, to maintain an
optimum temperature inside. Occasionally, however, white-
rumped swifts find their own nest sites in rock clefts. This,
I had been told, was the case at the castle.

'I have one!' says my German friend. 'Down there, over
the river, heading towards the white house.' With his directions I find and follow the fast-moving bird as it flies over
the woods and water far below, a dark, graceful crescent
with a patch of white. It is a fleeting glimpse, lasting perhaps
twenty seconds. I should perhaps be more excited at this
point, but twitchers' adrenaline does not flow through my
veins, glad though I am to have seen this other swift, to
know for myself that it is here. For a while I keep watching
and hoping for another view, but desire to explore beyond
this airy hilltop triumphs and I decide to take a walk along
the valley before the sun reaches its scorching zenith.

The path winds up and down little hills, through
scrubby cistus bushes. I find just a single late flower, white
petals splodged with purple; most were in bloom weeks
ago. As I brush past, the sticky leaves release their pungent,
medicinal fragrance. Hidden from sight in the streamside
vegetation, a nightingale utters its rich monologue of song,

the phrases almost conversational. A Sardinian warbler darts among the vegetation, unmistakable with its jet-black head and red eye ring.

A distinctive, conical hill beckons and it is worth the climb: a lookout shelter offers welcome shade. A group of friends, one of them carrying a toddler in a rucksack, are scanning the landscape through binoculars. They have spotted a black vulture sitting on its treetop nest about a mile away. The nest is huge, which makes it easier for us to locate. The pair return to the same tree each spring, adding more twigs every year, sometimes building a structure more than nine feet deep. Worldwide, numbers of this magnificent bird are in decline due to poisoning, habitat loss and changes in livestock rearing. Across Europe, in Portugal, France, Italy, Austria, Poland, Slovakia and Romania, they have become extinct. Thankfully, the Spanish population is rising, a direct result of conservation measures. Young birds from Monfragüe have been released in France in efforts to re-establish the species.

Later, as the temperature drops, I return to the hilltop castle. My German birding friend is there too, having spent four hours at Peña Falcon patiently hoping for more sightings of white-rumped swift. Finally, he saw three of them. His single-minded dedication is typical of the true birder; new birds must be learnt, not just ticked off.

As dusk falls, I walk along the track out of the village where I am staying, past a derelict football stadium with a high concrete wall and tattered, rusty wire netting. Few of the houses look occupied, most of them probably holiday homes. Pylons stand stark against the rose-gold sky, silhouettes of urbanisation. Then I hear sheep bells. A woman walks towards me with a bunch of weaver's broom, a shower of yellow flowers. The vegetation grows wilder, with more trees and thickets. Something is on the track ahead: wedge-shaped, tawny brown and with huge eyes. Softly, I approach. A nightjar! It flies and, as I watch, several more appear, rising from nearby bushes. They swoop and circle in the twilight, veering upwards then cruising into a glide as they hunt for moths.

With some reluctance I leave Monfragüe the next morning, after a final visual scouring of Peña Falcon. There is not a swift in sight so I watch the black storks, nesting on a rock ledge just above the river. One of the pair has just arrived back at the nest with a twig in its long red bill, which it prods and tweaks until it is woven into the structure. Extremadura holds 75 per cent of the Spanish breeding population of this bird – as well as the highest concentration of white storks.

It is time to find the other newcomer: little swift. Back on the coast in Cádiz, I arrive in Sanlúcar de Barrameda, a fine seaside town on the River Guadalquivir, from which Ferdinand Magellan set sail exactly 400 years previously, on a fateful voyage that accomplished the first circumnavigation of the world, but which resulted in his own death and that of the the majority of the fleet's crew. These days it is best known for horse racing along the beach and Manzanilla sherry.

Such delights would have to wait. I set out for a small town a few miles east of Sanlúcar. 'Go to the fish warehouse,' my friend had directed me. Swift-seeking does not always lead you to the most picturesque parts of town and this uncompromisingly functional, breeze-block building is unlikely ever to have featured in a tourist brochure. For those who love birds, though, it holds great charm. Within the metal girders of the loading bay roof are numerous old red-rumped swallow nests, now transformed unmistakably into little swift nests: stuck all over with feathers as if a child had been let loose with a pot of glue and told to make a collage.

And here too are the birds: house-martin-sized but unquestionably swifts, with scything wings, splitting the air with bursts of shrill twittering. They are paler than the pallid swifts among which they fly, their wings quite brown, a distinctive white patch wrapped around the rump and onto their sides. Their tail feathers are closed in flight, giving a square-ended appearance. This feature is quite unusual among swifts but is shared with the African spinetails.

All across their range little swifts have adapted to the urban environment and expanded alongside the rising human populations of Africa and India. In South Africa there is just one record from the nineteenth century; today they are common and widespread. Little swift has an aversion to forested areas so it is unsurprising they are doing so well now, with forests falling to the chainsaw all over the world. Urban habitats will never offer a fraction of the diversity of these ancient, natural habitats, but some species can thrive within them.

Now and then a little swift flies to its nest, perhaps to

take its turn incubating eggs, possibly to feed young chicks, though I hear no insistent calls for food. Then another bird flies up to one of the feathered nests – a pallid swift! It wriggles its way through the entrance, wing feathers trailing outside, and is then joined by its mate, though the tiny hole barely seems able to accommodate one of these birds, never mind a pair. Such is the urgency of finding a nesting place.

There are still just a handful of little swift colonies in Spain, all of them along the coast of Andalucia. Recent sightings in Portugal have raised hopes that they might settle and breed there too. Unlike the other swift species found in Spain, little swift quite often overwinters within its breeding range. Even in December and January they can be seen feeding over the Guadalquivir estuary.

On the opposite shore to Sanlúcar de Barrameda is the Coto Doñana, an internationally renowned wetland, with rich feeding opportunities for birds throughout the seasons. I take the ferry across, hoping to see marshes brimming with birds and humming with insects, but the wildlife-rich wetland I had expected to find had evaporated, leaving bare earth cracked as if with crazy paving, far into the distance. Rain had not fallen that spring, compounding problems caused by intensive agriculture. Large numbers of unauthorised wells and irrigation ponds have been identified in the area, sucking water from aquifers to sustain strawberry and arable crops. Meanwhile, pesticide runoff from agricultural activities is flowing into the wetland.

While parts of this precious nature reserve are still

thriving and providing insect feasts for birds, the loss of several extensive shallow lagoons and the majority of its marshes is of great concern. It is thrilling that humans have unintentionally provided an array of structures suitable for nesting swallows and the swifts that take advantage of them, but the drastic exploitation and pollution of the habitats they need for foraging is inevitably dispiriting.

21

SWITZERLAND'S ACTIVISTS

It is December and I have come to the shores of Lake Geneva in Switzerland. Christmas lights are twinkling, a golden haze of light; nothing like the flashing, multi-coloured, flaunt-it-to-your-neighbour displays in England. The swifts are far away, perhaps flying over rainforests in the Congo.

With its warm summers and vast lakes, Switzerland has excellent feeding opportunities for insect-hunting swifts and they breed here successfully – so long as they can find places to nest. But the problems affecting swifts in the UK and Italy are evident here, too: building restoration and reconstruction are continuously underway, and essential crannies, cracks and holes in the houses of villages, towns and cities are filled in and sealed off.

Fortunately, swift conservation is a growing force. The influence of the Swiss ornithologist, the legendary Emil

Weitnauer, lives on, thanks to his popular book, *Mein Vogel*, published in 1980. He knew swifts as 'Spyren', a local name that evokes phonetically their speed, verve and perhaps also their cry.

Fascinated from an early age, he remembered climbing up the village church tower and wedging himself under the beams of the bell frame to watch swifts on their nests. 'Time and time again I felt an urge to climb up to these wild, strange birds. I had no idea why. Something unique about them had taken hold of me. At that time I did not yet know that I would, for years to come – yes, more than forty years – be preoccupied with these common swifts.'

To make studying them easier, he fixed ten boxes under the eaves of his house. 'In the following years, during the short period when the swifts are with us, I was at their nests every day, with few exceptions. If anyone wanted me, they would find me at the swifts' nests.'

Weitnauer ringed his birds, and the extensive period of his study revealed fascinating evidence of the longevity of swifts. His book includes a photograph of one bird, taken on its twenty-first birthday. Over its lifetime it could have flown 6 million miles – almost a quarter of the distance from Earth to Venus.

The eager amateur famously took to the sky in a hot-air balloon to discover the secrets of the swift's evening ascents. He was thrilled by the whole experience:

We ascended from Bern at 11.30 pm on 6 July 1963 with a full moon, intending to be in the area of Zürich-

Kloten-Dübendorf at sunrise. A couple of times I saw swifts flitting by in front of the disc of the moon . . . we saw the sun rising over the Bodensee and a couple slid by under the balloon and more were flying by beneath us. The night flights were once again confirmed and this in itself was a great, unique experience for all who took part. Why should not the observers, like the swifts, also spend a night aloft without sleeping?

His pioneering work inspired David Lack to construct nest boxes in the tower in Oxford and influenced a succession of swift champions in Switzerland, including Roland Eggler, a station master who constructed and installed more than 2,000 nesting boxes, many of them placed under the deep eaves of railway buildings.

Today the baton of swift conservation has passed to others, including Marcel Jacquat and Bernard Genton. They too have written a book, *Martinet Noir: entre ciel et pierre* (*Swift: between sky and rock*), which I read with pleasure over many hours, with the help of a large dictionary. This visit would force my rusty A-level French back to life, albeit very imperfectly.

Marcel and Bernard met through bird ringing and have been friends for more than forty-five years; both had early careers as biology and environmental science teachers. Bernard later became a didactics professor for trainee teachers, while Marcel became the director of the Natural History Museum in La Chaux-de-Fonds in the Jura Mountains, where he has lived since 1970.

'We'll be in the sun in ten minutes!' says Bernard as

the road zig-zags through dense fog into the mountains. Sure enough, the thick cloud that is hiding the Alps from view suddenly clears.

La Chaux-de-Fonds is the heartland of Switzerland's watch-making industry, which originated in the sixteenth century, when jewellery was banned by religious reformers and craftsmen were forced to find alternative uses for their skills. More than a dozen companies started here; today it is still the world centre of traditional watch-making, with highly skilled craftspeople producing jewel-like time pieces for the luxury end of the market.

Marcel gives us a tour of this unusual city, much of it rebuilt after a catastrophic fire in 1794, its streets reconstructed afterwards on a grid system. Interestingly, it is the birthplace of Le Corbusier, renowned architect of urban planning. A streak of Art Nouveau runs through the buildings; opening a door from the street into an apartment block, Marcel points out an intricately carved plane tree and horse chestnut leaves over interior doorways and simply stylised flowers painted on walls.

And the city has swifts, breeding here at high altitude, more than 3,000 feet above sea level. Thanks to Marcel, La Chaux-de-Fonds is starting to be celebrated for its birds as well as expensive watches. He has made and installed almost 1,000 nest boxes under the city's eaves.

He builds them in his garage, tools neatly arranged across the walls, along with a few strings of onions, his car just fitting into the space in the middle. A neighbour lends him space in his garage, too, so he can store his nest boxes, ready for instant use. 'When you see scaffolding going up on a building where you know swifts are nesting, you have to act fast. Sometimes we put boxes up on the poles so that the swifts have somewhere to go while the work is carried out and, if permission is given, we will install boxes on the finished building.'

'A woman who lived nearby contacted me after she had read our book,' says Marcel. 'She told me her mother had been taught by Emil Weitnauer!' The schoolmaster's influence was still powerful; the woman asked Marcel if he would put swift boxes on her house. 'There are twenty-five of them now!'

As we walk through the streets, Marcel is greeted by many people, including the city architect and council leader. The architect, he tells me, is very sympathetic to swifts and is keen to incorporate nest bricks or boxes into building schemes. We look at a beautiful white-walled house with traditionally painted blue shutters; its swift boxes, tucked under the eaves, are painted the same shade of blue.

Integrating nesting space into buildings in an aesthetically pleasing way is something Marcel and Bernard do wherever possible. Sometimes the boxes are almost invisible, such as those installed between the cornices at the top of a church tower; often they are painted white, to merge with the colour of the wall. This also has the benefit of reflecting heat away from the boxes, so the swifts do not boil inside. A vineyard owner asked Bernard to make him nest boxes from wooden Bordeaux wine cases, a wonderfully individual installation.

While some people are excited about enticing swifts to their homes, Bernard and Marcel inevitably encounter obstacles, too. Alert to the destruction of a street of old houses, where a lively colony of swifts was known to breed, Bernard negotiated with the developer to put a rectangular, wooden swift tower on the roof of one of the sleek new blocks of flats built in their place. At the last minute, the owner decided against its installation, saying it would spoil the look of the building. Bernard was given permission to put the tower on top of a small electricity sub-station nearby, where it looks decidedly out of place and is unlikely to attract swifts, as it is very low.

He looks disheartened for a moment, thinking about this project and the swifts that lost their nesting spaces. Then he bounces back: 'In Geneva, the political authorities are on the verge of passing a new law that all new buildings over a certain height must incorporate swift bricks!' This is real progress, embedding provision of swifts into mainstream construction. Huge credit is due to Marcel, Bernard and Switzerland's other swift champions, who have raised awareness of the swift's requirements through numerous exemplary projects and publications.

Like all swift enthusiasts, Bernard has nest boxes under the roof of his own home. Fifty of them. An astounding thirty-five are occupied by swifts. Each summer he spends hundreds of hours watching their behaviour, noting the frequency of their visits, variations in calls and in flying techniques, learning all the time and discovering, too, how much he does not know.

Bernard Genton, installing swift boxes.

Other birds are encouraged to breed around his house as well: house martins, with a plank fixed under their nests to catch the droppings, and wrynecks, for whom a box has been slung low from an apple tree. These tiny, ant-eating woodpeckers were faring badly in the region; like swifts, they were suffering from a lack of nesting places. Then people started providing nest boxes – often seen nailed to trees along the boundaries of vineyards – and now their numbers are rising.

There is one bird, however, for which Bernard has little affection: the house sparrow. Like many swift lovers, he is fiercely protective of 'his' birds. Sparrows, of course, are also hole-nesters and, being resident, can take the pick of cavities long before the swifts return at the beginning of May. Except that Bernard and others have wily schemes to prevent this, such as fixing little doors over the holes, which are slid back with a long pole the moment the first swift is sighted.

This is not always possible, of course. High on a church tower, he points to a grisly spectacle: two desiccated swifts dangling from their nests, blowing in the breeze. The legs of these unfortunate birds had become tangled in string that sparrows had taken into the boxes as nesting material, almost certainly collected from a nearby farm-yard. Swifts, which only collect materials light enough to be wafted into the air, would never find string. 'Marcel says I shouldn't worry about the sparrows,' he says, 'but I do!'

Other people might be more concerned about the sparrows. Weitnauer's grim account of a pair of swifts that requisitioned an active sparrow nest demonstrates the absolute refusal of these birds to be evicted from a nesting site they have claimed as their own: 'In June, 1934, I saw swifts flying into a sparrows' nest at the schoolhouse, between the ridge beam and the gable end. It was a risky place to investigate, but it had to be done. To my great surprise, I found two young swifts on top of four dried-up young house sparrows, separated only by a thin, sticky layer of grass stems and feathers.'

Through his observations of colour-ringed birds, Bernard has noticed that older, more experienced swifts are better at fending off sparrows than younger ones. Stories are told of them sitting on nesting sparrows for hours at a time until finally the little brown birds give up and relinquish their nests. Swifts can look after themselves. And while they may encounter occasional problems, sparrows can also be inadvertently helpful. When looking for nesting holes, it is very often the chatter of sparrow chicks that alerts swifts to the whereabouts of cavities.

Many of the towns along the shores of Lake Geneva have a promenade along which you can saunter, looking across the water to the French Alps (when the fog clears). Plane trees grow here, shorn of all grace by repeated pollarding, but long-lived in recompense. These old trees are full of holes, perfect for tree-nesting ducks. Goosanders raise their young in them, undeterred by crowds of people passing by. As the old pollards die, people have started putting up nest boxes and the goosanders have readily adapted to them. Rafts of these elegant birds idle on the water, the males' heads a dark, shimmering green.

Bernard shows us around the tiny medieval city of Rolle, where he lived and worked for some years. A turreted castle next to the lake is now used for school rooms and council offices. As we walk into the mighty stone-walled courtyard, he excitedly describes the summer spectacle of swifts: 'They fly in a great circuit beyond and around the castle, their screams echoing off the walls! It is fabulous!' Very unusually, one pair even nests under the cloisters here, along with numerous swallows. 'The wife of the caretaker loves birds and so he is very tolerant of them,' says Bernard, happily. A swallow's nest can even be seen on top of a plastic movement sensor, along the open passage leading from the cloisters. 'We have put posters up to tell people about our swallows, so they understand about the mess. Sometimes we have as many as twenty pairs here – that is a big colony, for swallows!'

Few could fail to share his passionate delight.

Bernard's evocative description of the swifts' summer displays stays in my memory, so when I find myself in the Swiss Alps a summer or so later I seize the chance to return. During my days in the mountains I see a few common swifts speeding through the streets, tucking themselves into invisible holes in the pantiles of roofs on older buildings but, disappointingly, not a single living Alpine swift, though I find a stuffed, scraggy specimen in a museum. It gives me no joy.

Several trains later and a sweeping loop around Switzerland and I am back by Lake Geneva. Torrential rain is falling, the skies leaden and not a bird to be seen. Bernard wrings his hands in despair: 'The swifts have all gone to Lyon!' Of course they have. This is what the non-breeding swifts do when the weather turns horrible; they take off around the depression, often flying hundreds of miles to find somewhere warm, sunny and full of insects. From this part of Switzerland they simply follow the Rhône, where it flows out of the lake at Geneva, away from the mountains and into France, a distance of around 100 miles.

Bernard has a rainy-weather plan and heads into the hills to see a tower he has built for house martins. In compensation for extensive modernisation works in the village, he was asked to design some alternative nesting sites. It looks like an enormous, roofed bird-feeding table, but on the underside of the platform twenty-eight artificial nests are fixed, two rows on all four sides. It was installed just before the birds arrived, and with the help of a sound system playing house martin calls, attracted eight pairs in its first season.

We carry on to Montricher, a small village in the foothills of the Jura mountains, where we stop outside a house with an array of small tree-trunk sections under the eaves, imitating, perhaps, the original nesting habitat of the swift. These are horizontal but the principle is the same: a safe hole in something tree-like. The owner has made these nesting boxes not only for his own house, but for half a dozen others, too, persuading his neighbours that they will enjoy the pleasures of swifts.

All twenty nesting holes on his house are occupied and have webcams showing the activity inside, which he watches on screens in his garage. He scrolls through photos on his mobile phone, flicking past pictures of his grandchildren until he finds one of two young swifts with unusual markings: a white band across the belly.

Strolling through the village, we encounter another swift aficionado, a young man who also gets out his phone to show us something. This time it is a video of swifts in the nest box on his house. People really cherish and take pride in their avian offspring, the birds woven into the social fabric of the village itself.

The sky darkens as another rain cloud edges towards

us and Bernard takes us back to his house and supper, beautifully prepared by his wife Ariane. They have many fine paintings on the walls but Bernard directs my attention to a bell jar containing a swift skeleton, wing bones raised up, mounted on a metal rod. I look at the long, deep gape of the bird and I am amazed at its size. Just as a whale catches plankton, the swift opens its mouth wide, a vast container for the collection of hundreds of insects. I look at its short, stout humerus, the bone that connects it to the hummingbird; the common feature of both species through millions of years of evolution, working with the powerful pectoral muscles to provide different but supreme flying abilities in both birds.

Bernard and Ariane have a new project. They hope to attract Alpine swifts to nest in the ancient building in Montricher where Ariane works as a schoolteacher. 'There are more Alpine swifts nesting in buildings in towns now than there are in the Alps!' Bernard exclaims. I would have liked to have seen them in their wild stronghold in the mountains, where they nest on cliff ledges, but it is good to know they are expanding their territory in Switzerland. A few pairs are present in Lausanne, about eighteen miles away, so it is quite possible the birds will arrive.

Bernard is distraught that he cannot show me the evening swift drama that they have enjoyed every day for the last two months. Now with more than forty occupied nests on his house, the colony attracts scores of bangers, screaming and speeding over the garden. Even some of the parent birds may have disappeared for a few days to feed themselves. Their chicks will be fine; dropping in temperature by night to a state of torpor and warming up by day in readiness for their next meal.

Swift colony at Bernard Genton's home.

On my previous visit it was Mont Blanc that was invisible, though Bernard had excitedly pointed several times to where he assured me it was. Now, at the end of July, it's the swifts that are nowhere to be seen. There is no doubting him, though; his vivid descriptions are almost as entertaining as the swifts themselves.

The following morning is cloudless. Down by Rolle's handsome castle, each of its turreted corner towers slightly different from the others, swallows are swooping low over the lake and goosanders idling about on the blue-green water. People are busy setting out stalls and getting sound systems organised for the summer fête, young musicians gathering excitedly. There is, however, still not a swift to be seen. Only in my imagination can I see them swooping down to sip the water, shaking off the droplets that catch their wings. It is late in the season now and the non-breeders may keep going from southern France straight back to their wintering grounds; their migration journeys have no fixed dates.

22

NORTHERN IRELAND: INVISIBLE BOUNDARIES

Ireland is the common swift's most westerly breeding point, its presence as quintessential a part of summer here as anywhere in Europe. *Apus apus* has adapted to a great variety of climates across its vast range. Ireland is quite possibly the wettest, but offers an abundance of insects and long daylight hours in which to catch them, fulfilling one of the main requirements for the rearing of young. And while prolonged rain can bring problems for swifts, their legendary ability to circumnavigate weather depressions gives them the flexibility to escape the worst. They nest mostly in urban areas, all over the country, avoiding only a few places in the most exposed western fringes, blasted by Atlantic rainstorms. Occasionally, in the coldest, rainiest summers, they may abandon their nests and return to Africa, a strategy that gives the adults a better chance of survival and an opportunity to breed the following year. Some people wonder why they come here at all, why they do not simply stay in Africa all year round. The answer, of course, is that it works for them and has done for thousands of years.

This watery land has more than 12,000 lakes, the largest being Lough Neagh. Rain from almost half the surface of Northern Ireland drains into streams and rivers that flow into the lough, including the rivers Maine, Six Mile Water, Upper Bann, Blackwater, Ballinderry and Moyola. Twenty-two miles long and twelve miles wide, this freshwater lake in County Antrim is a wonder to behold – when it is visible. Strangely, the lough seems as elusive as Mont Blanc on a foggy day, for there is no shoreside road. Birds, of course, find it easily; this is the biggest wintering site for diving ducks in the whole of Britain and Ireland and in summer it is a magnet for swallows, martins and swifts, for whom Lough Neagh provides a feast of flying food.

Clouds of chironomid midges emerge from the lake during the summer months, ghostly insects with grey wings that thrive in the humid conditions. A swarm of these flies is almost as remarkable as a starling murmuration, billowing like smoke above the shores of the lake. Mercifully, though, the Lough Neagh midge does not bite – it has no mouth. It emerges from the water to mate, lay eggs on the surface of the lake, and then it dies, drifting down to window sills in dense layers, falling like snow into people's hair, or snapped up alive by fish, dragonflies and insectivorous birds. Its lifecycle takes just two or three weeks to complete in summer, providing a continuous abundance of food for its predators.

It is a couple of days after the end of the marching season when I arrive in Northern Ireland, but Union flags and those of the Orange Order are still draped over house fronts and flapping from metal arches rigged up over streets. Burnt-out bonfire circles the size of merry-go-rounds, with here and there the charred remains of mattress springs, are all that is left of the towering pallet-stack piles that are torched annually on the Eleventh Night, the day before the Twelfth July Protestant cele-brations. The parades passed off peacefully enough, though two houses had burnt down, accidentally set on fire by flying embers.

Half a mile from the north-east shore of Lough Neagh is Antrim town. I have come here to meet Mark Smyth, a man as fanatical about swifts as many of his countrymen are about their 400-year-old marching traditions. 'You will see two silver birch trees and three dodgy-looking blokes outside my house,' he told me in his directions. The suspicious characters turn out to be respectable scientists from the RSPB, holding up a long black net. They are hoping to find out precisely where local breeding swifts are feeding and have come to catch the bird they fitted with a GPS tag ten days earlier, a breeding adult with chicks, in one of the nest boxes fixed to a wall of Mark's home. The swift, however, is having none of it, staying quietly in its box, occasionally looking out and then retreating into its dark hole.

Overhead, the grey sky seems to be full of swifts, slicing over the rooftops, gliding higher and feeding. 'That one's got a full crop, it'll come and feed its chicks in a minute,' says Mark, pointing to a bird which looks exactly like all the others. 'Look! You can see the rounded shape, it's different.' I watch intently and, yes, one of the birds is plump at the throat, I can see it now, the bolus of insects distending its gorge, a meal ready for delivery.

While we wait, Mark tells me how his passion for swifts started. 'I noticed them nesting in a hole under the eaves of a house at the back. The owner got the board repaired and blocked up the hole, so I started putting up boxes on my house and playing swift calls to entice them over.' High on the pebbledash walls, on the front and the gable end, he has rows of them, twenty boxes in all, with swathes of black cable looped up to his bedroom window, linking webcams installed in each box to his computer.

With ten boxes occupied, Mark has excellent opportunities to learn about swift behaviour and, when not at his job at Asda, he is constantly checking and monitoring their activity. He keeps detailed records of the swifts' activity: the timings of eggs laid, hatchings, chick development, frequency of feeding by parents, and ultimately of their fledging. In these confined spaces the birds inevitably spend a lot of time quite still – incubating, sleeping and preening – but there are moments of drama. The speed at which swifts approach their nests is breathtaking for us to watch – and occasionally also for the bird: Mark has footage of a swift shooting into its box so fast it hit its head on the top of the box and fell down.

Watching closely, he catches tiny incidents in the swifts' lives: 'While sitting in my garden I saw one of my swifts make a few attempts to line up with the gable boxes. It was carrying a large white feather which made it look like it had a huge patch of white on its breast. Other swifts kept chasing it as if they were trying to get the feather for their own nest.'

Quite often he witnesses fights. Usually these happen when an intruder enters the nest box, but sometimes they occur between pairs. Battles between mates can lead to egg dumping and then the whole breeding process begins again. 'Why do they do it? Anger? Adrenaline? Does the male suspect his mate has been unfaithful?' No one knows.

For Mark and others like him who watch their swifts closely, such behaviour raises questions about whether they should intervene or simply let nature take its course. Given that they are not taking part in strictly controlled scientific experiments and their efforts are geared towards conservation, they will occasionally take action if a bird's life is endangered. Just a few weeks ago, Mark rescued a bird caught at the entrance hole, in the grip of its rival's claws. Given that swift fights go through prolonged periods of deadlock, this bird, which was hanging outside, was extremely vulnerable to passing predators.

Meanwhile, the RSPB men are still waiting for their bird. And then it is there, shuffling out to the entrance then up into the air. The two men flick the net over and catch the bird in its folds. This is the first time I have witnessed flick-netting, a controversial technique. Swift

Conservation states that swifts should not be trapped in any form of flick net: 'Swifts are not tennis balls. The likelihood of injury to their highly developed and refined wing and bone structure is very great. Newly ringed swifts have been recovered injured or dead.' There have been reports of ringers catching large numbers of swifts in early May using flick nets; on overcast days the birds will fly low to hunt insects. There are concerns about the stress inflicted on them at this particularly vulnerable moment, when they need time to feed and recover from their long migration journey. Swift Conservation strongly advises against ringing swifts on their arrival. The British Trust for Ornithology guidelines are less stringent.

Most ringers are scrupulous about the welfare of the birds they handle, however, and the operation I watch is carried out quickly and efficiently in good weather. The tiny GPS tag weighs less than a gram. It has a minute aerial, about an inch long, which picks up location details, giving a record of where the bird has been feeding.

Kendrew Colhoun quickly cuts and removes the thread-like harness and logger from the swift's back. He raises the bird in the air on the open palm of his hand and straightaway it is off in a loop and up in the air to join the other swifts racing overhead, seemingly untroubled by its experience.

'Right – let's go to Belfast and meet Peter!' suggests Mark.

In a city-centre café we meet Peter Cush, a scientist recently retired from Northern Ireland's Environment Agency, roughly equivalent to England's Defra, the government department dealing with agriculture and other rural matters. Around the table a lively discussion erupts about the merits of these experiments on swifts. Peter questions the necessity of the RSPB's trial: 'We need to ask ourselves – why are we doing this? We know where your swifts are feeding, Mark – Lough Neagh!'

Early results from the RSPB's GPS analysis have indeed confirmed this. Peter is concerned about the effects of catching the birds and how the experience affects normal feeding behaviour. There are indications that the process of having the harness attached – which takes about half an hour – is causing some of the birds stress, leading to uncharacteristic behaviour and even desertion. Two of Mark's birds were gone for ten hours, one for twenty-six hours, and one, which was carrying a geolocator, failed to return at all. Perhaps it will be back next year, with a full record of its journey.

Peter elaborates:

We need to think very carefully what exactly we are trying to find out before undertaking this kind of research because swifts are highly sensitive to handling; it can cause them to abandon their nests. There should be minimal disturbance – imagine how it is to be a swift and you might be able to work out what should and shouldn't be done. Their nest should feel like a place of absolute safety. You need to ask yourself how useful the information will be – is it telling you something new and scientifically worthwhile that justifies the risks to the birds?

With swifts in decline, there is anxiety about every bird. New scientific technology is exciting but must be used with care. The weight of the latest devices is minimal, but attached with a harness to the bird's back, they slightly disrupt the perfect streamlining of the swift's body. The main concern, however, is catching the birds and fitting the harness.

Migration trials involve fitting the device at the end of the breeding season, so avoiding the threat of desertion, but research aiming to discover feeding behaviour during the breeding period is more invasive. However, the number of birds involved in such scientific trials is very few and unlikely to impact on population numbers. Nevertheless, it must always be asked whether the research will be of benefit to the birds, or simply to the scientist's career.

Over the next few years, however, the survey yields some interesting information. Confirmation that birds close to Lough Neagh hunt for their food over its insect-rich waters was hardly surprising, but data gathered from breeding birds in Belfast showing that they too flew back and forth to the lough – a fifty-mile round trip, repeated multiple times each day in order to forage for their nestlings – provided evidence that these urban swifts are having to work very hard to find their food. Intensive farming systems, seen in much of the countryside surrounding the city, are not conducive to thriving insect populations.

We walk along the streets of Belfast, stopping outside a stone-built Victorian building close to Queen's University. This former girls' school is now the Crescent

Arts Centre, a vibrant hub for cross-community events, where people from all sections of the city come together to sing, dance, write poetry and explore the visual arts. It is also home to a large colony of swifts, which nest under the eaves. This four-storey building underwent a major programme of restoration in 2009 and Peter worked closely with the builders and architects to ensure the nesting holes were retained. In addition, people sponsored swift bricks, which were installed in an extension to the side. As we watch, a swift swoops into one, while dozens more circle the building. Their crescent-shaped silhouettes have become living emblems of the arts centre and its aspirations.

Mark and Peter have spent many days this summer searching for natural swift nest sites in Northern Ireland. Someone reported seeing a swift dart into a crack in the rocky cliffs on the Antrim coast and they decided to follow up the report. Two nesting holes were found. That evening we take the 40-mile journey to Larrybane to see them. A long way, you might think, to look at a crack in a rock. The route takes us via the Dark Hedges, an avenue of beeches planted near the village of Stranocum 250 years ago. The trees' branches reach across the quiet road, twining into a Gothic ogee-shaped tunnel that has become famous as a film location for *Game of Thrones*.

It has been drizzling most of the day, grey clouds louring over the lush pastures. As we arrive at Larrybane around nine in the evening, the rain clouds are finally blowing away, brightness gleaming over the sea. We walk down through a vast amphitheatre in the old limestone quarry. Purple marsh orchids are flowering and Mark scours the ground, looking for an unusual hybrid he has seen before. A short scramble down a rocky path to the shore and we come to a recess in the white chalk cliffs, flints embedded in the boulders. At the time these rocks took shape, perhaps 100 million years ago, birdlife was evolving rapidly; some of them had teeth and clawed fingers at the tips of their wings.

The swifts we have come to see have a different lineage and their own mind-blowing history. Pointing out the two narrow crevices from which he has seen swifts emerge, Mark speculates that they may have nested in these cliffs before people settled on this island.

Turning around, we find the sun streaming onto the cliffs around the bay, soaking them in a rose-gold glow. Across the water, Rathlin Island, barely visible ten minutes ago, is lit up and a rainbow rises from the sea, arching over Carrick-a-Rede, a rocky island with an enormous colony of razorbills and guillemots. It is just twelve miles from here to the Mull of Kintyre in Scotland. A few minutes later and the sun has dipped, twilight turning the sea a deep storm-blue.

As the light fades, a swift appears from a curve in the cliffs. It flies up and away from the coast, perhaps heading for a midgy lough. Last time Mark was here he saw a screaming party of seventeen, and it is highly likely there are more pairs breeding here.

The next morning we return to Antrim, finding Mark in his garden, a swirling sea of cranesbill geraniums in shades of blue, purple and pink. An avid gardener, he has a particular fondness for snowdrops: 300 varieties lie dormant underground; Mark never does anything by halves.

Today we are joined by James O'Neill, a young zoology student at Queen's University with a passion and curiosity for all kinds of wildlife. He lives in Portadown, near the southern shore of Lough Neagh, in Armagh. James had not planned to spend the summer at home. After finishing his dissertation on the cryptic wood white butterfly, a species recognised as distinct from the wood white only in 2011, he set off for Romania where he had landed an eight-week job leading an Operation Wallacea conservation research project on butterflies. Unfortunately, he had tripped on a steep slope in the Carpathian Mountains, breaking his collarbone, and was forced to return home.

Portadown is notorious for the violent strife that broke out between its Catholic and Protestant communities during the Troubles. Located in the so-called 'murder triangle', more than forty people were killed here by Loyalists and Irish Republicans, the majority of them civilians. The annual orange parades often led to riots and shootings and continue to be a source of tension. During peace talks in 1998, a massive car bomb exploded in the town centre, threatening to derail the process. The wounds of those years will take generations to heal. As the metal barriers that divided sections of the town are gradually removed or given new purpose, some residents take pride in the enormous population of swifts, nesting under the eaves of houses inhabited by Catholics, Protestants and non-believers alike. These birds have their own territory and sometimes bitter struggles too, but away from their nesting places they mingle in the skies overhead, hunting and feeding above the rooftops.

A year or two ago James spent each evening of June and July seeking out their nest sites and recorded 135. He shows us around. I spot a swift coming out of a hole by a drainpipe above a hairdresser's. This is a new one for his list, but he is not surprised – the town is screaming with swifts. He takes us to the site of a demolished pottery factory with piles of rubble on the ground and a long, high brick wall down one side. Swifts and sparrows are nesting in holes in the brickwork and it is not long before one arrives. Amazingly, it does something swifts rarely do: it passes its bolus of food to the chick while hanging at the entrance, clinging to the brickwork with hooked claws. The young are growing, and there is so little space in this nesting hole that the parent bird has to stay outside while feeding its young.

Not every town in Northern Ireland is so happily populated with swifts. These birds were sometimes caught up in the Troubles too, little-noticed casualties of destruction. In Newtownards, County Down, a huge bomb exploded on 5 July 1983, thankfully without human fatalities but lifting off many town-centre roofs and wiping out in an instant much of the local swift population. Meanwhile, the familiar problems associated with overzealous renovation and contemporary building materials are further reducing the availability of nesting holes in towns and cities and numbers have inevitably dwindled. A 25 per cent decline has been recorded over the last quarter century.

Our journey today takes us to the rural county of Fermanagh, in search of more tiny crevices in rocks. As we drive, James tells us that his father remembers the call of the corncrake from his childhood. This secretive bird nests on the ground, and the change from haymaking to silage, with its earlier mowing dates, has virtually wiped out the breeding population across the UK, with the exception of the Outer Hebrides and Orkney. For younger generations, the corncrake is a rare bird that they may seek out if determined and ready to travel, but fifty years ago it was familiar throughout the Irish countryside, a regular summer visitor with a rasping night-time call that might sometimes have disturbed their sleep but whose disappearance was a matter of deep regret. Now virtually extinct in Northern Ireland, it survives in a handful of places in the Republic of Ireland. What strange, lost birds will the next generation be telling their children about, I wonder. The curlew with its hauntingly beautiful call and beak curved like the new moon? The whoop of a lapwing as it tumbled through the skies in joyous springtime displays?

James is a trainee bird ringer and he enjoys the chance it offers to get close to birds. Some ringing opportunities occur unplanned. Ever watchful of what is around him, he was walking through Belfast after an exam at the end of term, when he spotted a gull harrying a bird of prey. Never without his binoculars – even when going to an exam – he identified the bird as a young peregrine falcon. It was struggling to fend off the gull. He followed them and saw the peregrine drop into a garden, dazed but uninjured. Wary of its powerful yellow talons, he picked it up, took it home, ringed it and sent the record off to the British Trust for Ornithology. After a few hours of quiet and darkness the bird recovered and he set it free.

Like nearly all those I meet on my swift travels, James has nest boxes on his house. He put them up last year and straightaway a pair moved in, breeding that same season. Perhaps they had been displaced from another hole. He talks with excitement about how he saw one of the chicks about to fledge: 'It was the evening and I saw its white face peering out of the box. I was fearful and wanted it to go back but it kept coming, then it opened its wings and crash-landed in the bushes. One of its parents appeared behind it and the fledgling managed to get airborne. They flew around in a circle together and were off.'

Driving through rolling farmland, we pass a line of silage bales, graffiti-painted as a giant caterpillar, each bale a separate segment of its body. Then we meet John and Georgina Young, two dedicated swift enthusiasts who are creating wildlife heaven around their house, including a large pond, brimming with life. Swifts and swallows swoop down to drink there in the evenings. John has a Midas touch with swifts. There were none in the area when he put up his first box, but soon after he started playing the calls they appeared and now he has a colony of ten pairs.

Together we take the short journey to Carrickreagh Quarry, stopping on a quiet lane among fields. Trees have sprung up around the edges since industrial activity finished, but the quarry floor is marshy and pinging with tiny frogs, and four-spotted chaser dragonflies rustle through the air. Watching for swifts leaving and entering the cliff face requires serious patience. We have our eyes fixed on the cliffs but I am talking to Georgina about Kelly Kettles. Shaped like a large Thermos flask, but hollow down the middle so you can light a fire inside, these steel camping kettles are brilliant. Keep one in the car with a handful of dry sticks and you can brew up a mug of tea in minutes. Georgina says she must get one. Mark orders us to stop chatting and concentrate.

John calls out and we catch sight of a swift flying from the cliff. Mark, however, has missed it and looks outraged; not much gets past him. Within an hour five more nest sites are identified. I think of the swifts nesting in the limestone rocks on the Shropshire–Wales border. There may well be many more swifts breeding in cliffs and quarries across the UK than are recorded. So much of it is about being in the right place at the right time to catch that first glimpse of a bird diving into a crevice, then spending hours watching and waiting to see if there are others.

As we leave, James pauses beneath a willow tree and finds the chrysalis of a lunar hornet moth. By mimicking the appearance of hornets these insects have evolved a powerful defence against predators. Their larvae feed in the roots and trunks of isolated trees for two years before emerging with yellow-and-black-striped bodies and clear wings.

Back in the car, he entertains us with bird sounds. He is a fine mimic with a repertoire that includes heron, curlew, reed warbler and snipe. Once, while making tawny owl calls, a young owlet landed on his shoulder. He tries to teach us how to make cuckoo calls, but only the sound of puffing comes as we blow through our hands. We pass through Dungannon, which has a football team called the Swifts; they do not know their birds, though, as their emblem is unmistakably a swallow.

On our return, we finally get to see Lough Neagh. It is spectacular: an inland, freshwater sea. Its sparkling waters are not, however, in good ecological condition and both fish and birdlife have declined. The rivers that feed the lough are loaded with agricultural phosphates and polluting urban runoff, causing eutrophication. One creature, however, continues to thrive in these high-nutrient waters: the Lough Neagh fly. Thank heavens for the swifts and hirundines that snap up millions of them each summer.

On a peninsula jutting into the lake from its southern shore is Oxford Island Nature Reserve. At its tip is a visitor centre with a whole shanty town of house martin nests under its eaves. Many of the mud cups are built adjoining each other, more than 100, clustered in crowded intimacy. Each has its own entrance, mostly at the side. The July air is full of martins – adults and first brood birds; a willow tree on the edge of the lough flickers with wings as they alight for a few moments, chattering to each other.

Close to the eastern shore of Lough Neagh is Brian and Deirdre Cahalane's house, where we have all been invited for cheese and wine and the chance to enjoy the evening swift action. First, Brian declares, we must go to the local Tesco supermarket. This is no shopping trip, however. In

the car park there is a white wooden box shaped like a
dovecote on a pole. It is, of course, a swift tower. A sparrow
pokes its head out of a hole. 'They haven't been playing
the swift calls,' he says, frustration edging into his voice.
A starling squeezes itself out of another hole. These bird
towers have their uses, if not for the intended species, but
are proving a dubious way of going about swift conservation.

There are many cheaper and more effective ways to
help swifts, yet towers are gaining popularity around the

Mark Smyth, Brian Cahalane and Peter Cush.

UK. But without follow-up and playing the calls, swift towers can become something of a box-ticking exercise to demonstrate positive action for the environment. Mark talks about one erected on a nature reserve by Belfast harbour, which cost £40,000. 'With that money you could buy 800 Schwegler swift boxes, 1,142 Filcris recycled plastic swift boxes or 2,666 of John Stimpson's wooden swift boxes!'

The route back takes us close to the shore of the lough, but once again the water is out of sight, views blocked by newly built houses. 'Do you not have a planning system here?' I ask. Peter Cush shrugs his shoulders and mutters his reply with world-weary scepticism.

The swifts, though, are not bothered by the intrusive buildings; they might even find them useful in fifty years or so, if cracks start to appear. There is a cloud of them flying above us, whirling, racing and snapping up the bountiful flies.

Brian's house is a living wonder. He shares his brick-built modern home with more than 100 birds: thirty-five pairs of swifts, fourteen pairs of house martins and dozens of their offspring, snuggled inside their nests. It is often said that swifts and martins do not co-exist harmoniously, but the evidence here contradicts received opinion. One pair of martins has even built their mud-nest cup in a tight space between two swift boxes. Beyond the house, in the garage, there are nesting swallows.

After supper, we sit outside and watch swifts. Some are in groups, racing their aerial circuits, screaming their air-splitting calls. Others are by themselves. A singleton comes by, giving three short bursts of its call. 'It's saying – "I have a nest!" It is shouting for a mate,' says Mark.

There is one strange thing about this swift colony. Each year some 20 per cent of the fledglings fail to get airborne on their flight from the nests. This is odd, as the house is tall, with a long drop to the ground. This year has been the worst ever, and Brian has picked up twenty-five young birds. He is especially watchful at this time of year, fearing what may happen. When he finds a crashed bird he puts it gently into a wooden box and keeps it till dusk, giving it time to recover. Then he takes it down to a field near Lough Neagh, retrieves a hidden stepladder, climbs up and lets it go from the palm of his hand. All have flown successfully, but it is an anxious time for him when the birds are fledging. Such a high rate of first flight failure is very unusual.

Relaxing in a deckchair, waiting for the swifts to return, I wonder at the passionate dedication of people like Brian, who strive so hard for these birds. Such obsession is never going to be matched by the population in general. Nature has been banished to the margins of our lives, excluded from ever greater areas of the world, but there is growing recognition, at last, that things must change. We have reached a tipping point and the need for action is palpably urgent. It is not necessary for us all to do the kind of things they have, but it is essential that we care and follow that through with actions that give nature a better chance. The achievements of these swift enthusiasts are inspirational.

Light is ebbing from the twilight sky and all at once the swifts are flying into their boxes. Some still have a nest of chicks, others are breeders whose young have fledged and gone, staying on a few days to renew their energy levels before heading south. The not-yet-breeding two- and three-year-olds will spiral up into the sky and sleep on the wing.

23

LATE SUMMER

The brief span of time that swifts are present makes late July seem like the end of summer. Or perhaps just the beginning of the end. Swallows and house martins will be among us, fingers crossed, for another two months, but our swifts will be gone by mid-August. I want to make the most of their last few weeks here, store up memories of their wild exuberance, to cherish in the months ahead.

So I go to a place where I know they will be present in strength: an old workhouse, just across the border in Wales. Half a mile or so out of the small market town of Llanfyllin, it is set in a wide, green valley near the Berwyn Mountains. Once a grim institution for abandoned children, unmarried mothers and the destitute, the workhouse is now run as a vibrant community venture, with artists' studios, a museum gallery and an annual music festival. Its crumbling masonry is full of swifts.

It is an impressive building but austere. Constructed in the style set out by the Poor Law Commissioners, it is built of limestone in a cross formation, with four wings radiating from a central octagonal tower. From here, the master of the workhouse could view each courtyard and the comings and goings of the 250 occupants. A strong resemblance to prison architecture is no accident; it was intended to strike terror into the feckless and idle, and spur them on to work for their living. Opened in 1840, it rapidly filled up. Few of the residents were able-bodied, the majority children along with unmarried mothers, the elderly and infirm. At the age of two, children would be separated from their mothers, with men, women, boys and girls occupying separate quarters.

'Workhouse' meant what it said, and inmates spent long hours labouring. This included stone-breaking, grinding corn and the tedious task of picking oakum, unravelling the fibres of old ropes for use in caulking ships' timbers. At the age of twelve, girls were sent out into domestic service and boys were hired out for farm work. As the nineteenth century drew to a close, workhouses increasingly became sanctuaries for the elderly and sick, and in 1929 legislation was passed to allow local author-ities to take over workhouse infirmaries as municipal hospitals. But it was not until the National Assistance Act of 1948 that the last vestiges of the Poor Law disappeared, and with them the workhouses.

Renamed Y Dolydd (The Meadows) in 1920, it morphed easily into an old people's home, though mothers with children were still living there as recently as 1969, and homeless people were being given lodging for the night into the 1970s. By then, Y Dolydd had been modernised somewhat and had outgrown its grim reputation to become a much-loved refuge. When its future was threatened, a passionate campaign was fought to save it, resulting in a twelve-month reprieve. Financial pressures finally forced its closure in 1982.

Many former workhouses have been demolished or converted into private accommodation, but after a period of uncertainty, when it passed into the hands of property developers who rapaciously asset-stripped it of fittings, fireplaces and floorboards, Y Dolydd found a new purpose as a community centre for the arts, music and education. A strong environmental ethos underpins its work, and this is given living expression by the large colony of swifts nesting around the building. An information leaflet about its history is illustrated with a striking wood engraving of swifts racing over the building. 'This place belongs to the swifts,' says a woman working voluntarily in the garden. 'It feels as though we are caretaking the building for them.'

That hot July evening I witness the most breathtaking swift display of the summer. Impossible to count, there are at the very least seventy birds flying overhead, gliding, flickering, chasing. This is what I had hoped to see in Switzerland. To see them in Wales, so close to home, is even more exciting. I enter the first courtyard and instantly a squadron of low-flying birds tears across and sheers over the roof opposite, their piercing screams bouncing off the walls. Barely a moment passes without at least one swift flying up to the eaves, immatures clinging briefly to the walls, scouting for nests, and parent birds arriving with perhaps their tenth meal of the day.

Swifts at Y Dolydd, engraving by Bob Guy.

Red-gold light streams through the clouds, the sun sinking fast. Then bird after bird dives under the roof and into holes in the stonework, two or three pairs on each side of the courtyard. This is just one quarter of the building; there are many more in the other sections and around the outer walls.

Did swifts nest here when the building was used as a workhouse? Probably – its open eaves would have given the birds access since it was built. Men, women, boys and girls may have watched these birds from their separate quarters, wondering at and envying their freedom to go where they could not. The elderly residents of Y Dolydd would have watched them, too. No longer active themselves, some, at least, would have revelled in the birds' aerial mastery.

Gradually, the building is being restored, but work is proceeding with its occupying swifts and five species of bat in mind. It is quite possible to retain crevices in the stonework while repairing the stonework.

Back home and I am clattering down the narrow passage next to my house, wheeling out the bin. I hear something above and stop. It is coming from the eaves of the house next door: the eager noise of nestling swifts trying to attract a parent bird's attention. I treasure these silvery sounds. Still, not a single swift has shown any interest in the nest boxes installed on my house seven years ago, and the only response I have had to the swift calls blasting out from the window closest to the boxes has been from my daughter, who thinks this non-stop screeching is simply my odd choice in music. One day, though, these sturdy boxes will be used by swifts, sparrows or bees.

Standing in my front garden one evening, enjoying the wild races of the bangers, I see a swift pause at the highest point of the eaves of my house above the passage. It clings to the brickwork for a second and is gone. Another circuit and it is back. And again. Immediately, I want to do something, to provide the dark hole this bird is so clearly looking for. A box is out of the question on this side of the house, though; the height of the wall and the narrowness of the space on which a ladder could be placed forbid it. Maybe I could knock a brick out of the wall at its apex from the inside, I wonder. Fortunately, perhaps, I am advised that this would be unwise.

A week or so later I inspect the attic, a place I generally only go to retrieve camping equipment or Christmas decorations. I had not previously explored the hidden recesses on the side of the house the swifts had shown an interest in. Shining a torch onto the wall, my eye is caught by faint whitewash markings. They have been here before! Flicking the light beam higher, the cause of the swift's inability to get into the building is made visible. Pieces of blue plastic have been shoved into the tiny gap, along with shots of expanding foam, blocking off their access. It takes just a few minutes to remove the wretched stuff. Not only is there now a chance I might have swifts under my roof, I also have a well-ventilated roof space.

In its last week as a nestling the young swift sits near its entrance hole, looking out for most of the day, preparing itself for its momentous launch into the aerial world. Occasionally, it retreats into the darkness to exercise its wings and receive food from its parents, but always it returns to the entrance, as though pulled by a magnet.

An inexorable instinct to fly powers them out, but they are also encouraged by other swifts. Derek Bromhall, filming in Oxford, sensed an air of excitement around the tower at this time. Morning screaming parties of non-breeders screeched in ever-greater exhilaration, flying close to the nesting holes where the young birds were peering out into the world. Sometimes one would even drop down to the roof for a few seconds before launching itself into the air, back to its companions.

Caught between fear and longing, the young bird often makes several attempts to leave the nest before its first flight. Close to the edge, it tips itself forwards, panics and scrambles back to safety. Eventually, after a few minutes, hours or a day, it will tumble out, drop and fly. From this moment it will be continuously on the wing, day and night, until such time as it finds its own nesting hole a year or two later.

Fledgling swifts leave the nest while their parents are out searching for their next meal. On their return, a large ball of food in their throats to feed their young, the adults are faced with an empty nest and seem visibly perturbed, just as we might, by the absence of chicks. They will poke the nest with their bill, as if to check there really is nothing there, and jump on and off it. Meanwhile, the fledglings, untaught and impelled simply by instinct, head almost immediately for Africa – often the same day – where they will wander the skies for the next ten months.

A trickle of other birds may head south, too: a few swallows, willow warblers and sand martins, though the majority of these will leave later. Cuckoos are already long gone. At the same time as the migrant birds are departing, southerly winds are wafting painted lady, clouded yellow and red admiral butterflies over the sea to our shores, along with immigrant moths, such as the wonderfully named splendid brocade and silver Y. Rosebay willowherb, or fireweed as I know it, glows along railway lines, and rowan berries are turning a dull orange, but summer is still here. Every day brings something different in the ceaseless continuum of the year's cycle.

While it is common for swifts to renew mating activity once their young depart, the restive, migratory instinct almost always overpowers the reproductive urge. While many of the birds that nest among us, such as blackbirds, robins and swallows, often raise several broods of young, it is extremely rare for swifts to do this if the first clutch is successfully reared.

However, a curious phenomenon occurred in Germany in 2004. No fewer than thirteen out of the fifty-four pairs nesting in three buildings in a small town north of Frankfurt refeathered their nests and laid a second clutch of eggs. These included one pair which had successfully raised four chicks. The principal colony was at the house of Erich Kaiser, meticulous recorder of swift behaviour since the 1960s. While eight of the clutches were abandoned after several weeks of incubation, chicks hatched from the remainder. At least one of these resulted in fledged birds. The migratory instinct belatedly overtook several other pairs; their chicks were then rescued, hand-reared and released. Observers elsewhere also noticed this extraordinary behaviour in 2004. Though unusual, it is possible that second broods occur more commonly than previously thought. It is, after all, only recently that people have watched and monitored swifts so intensively.

Yet what was it in that particular year that influenced their behaviour? Superb weather? A massive abundance of food? Kaiser could find no explanation in what had been, in other ways, a generally unremarkable year for swifts.

It has been a fine summer with good feeding opportunities and swifts are fledging relatively early. I think the young birds next door have flown; I no longer hear their calls. Sitting in the garden at dusk, I wonder if the parent swifts are still around. I watch the evening primroses perform their magic, slowly unfurling tightly rolled, umbrella-like petals, which snap suddenly into luminous, brimstone flowers like mini-satellite dishes. Moths will find them within a few minutes, drawn by their scent and glow. Just as I am losing hope of the swifts' return, they swoop by within a split second of each other. The leading bird does its characteristic handbrake turn to negotiate the narrow space between the houses and disappears under the eaves. Its mate, hard on its tail feathers, flicks up and over the roof, circling round the houses before entering. Three evenings in a row I see the same thing, the pair's bond evident in their synchronised flight. Watching these high brick walls can bring moments of unexpected pleasure. On the fourth night I keep looking as the light drains from the sky. Then a bat appears and I know they are gone.

Numbers are dwindling but a few are still swirling around the skies. Then the rain comes. Lying in bed, I listen to it; this is one of my favourite sounds, refreshing and soothing. Soon I am drifting off to sleep.

The next day the water barrel is brimming so I top up

the pond and rescue the purple loosestrife, its leaves crisping at the edges. Another deluge arrives in the afternoon, driven by a gusty wind. I do not see a swift all day. As the rain eases off I take a walk in the hills. The turf is speckled with the yellow of tormentil, and there are tiny white eyebright flowers, scattered like raindrops.

Half a rainbow appears, a pillar of colour, shifting in the distance. Back home and the sky is amethyst grey, the rainbow complete now, arching over the street. The garden is lit with low beams of sunlight, every leaf and flower glowing with intense colour. Wisps of pale gold clouds drift over the rain clouds. And then the swifts appear, a loose gathering gliding and flickering through the rainbow.

My local birds may have disappeared, their piercing calls absent from the streets, but over the next few weeks I still catch sight of swifts if I look up at the right moment, circling overhead as they feed on fine evenings or head steadily south. I can hear them too, their cries mellowed by distance; these birds are flying high. By the third week of August they have almost all departed. Only a handful of late nesters remain now, attempting to raise their brood before it is too late.

At this point, swift-watchers are often overcome by a wave of melancholy. No more swifts for 250 days! Summer's over! Tough old bean that I am, this is not quite how I experience their departure. Which is not to say that I do not miss them. After three months of their ringing calls, singly or in gatherings, the sky seems oddly quiet and the birds that are left remarkably slow. And when a south-bound group of them appears a couple of days after I thought they had all gone, I am as thrilled as ever.

But I also love that they go, that their lives are lived elsewhere for nine months of the year, without a shred of dependence on humankind, roaming over rainforests and savannah, always in the air. For ninety days northern Europe's lengthy daylight hours and relative abundance of nesting holes have been useful to them, as they have for thousands of years. Then, two or three young raised and job done, they return to a totally aerial existence. It is their otherness that makes them so fascinating. They touch our lives briefly and then vanish; this is part of their magic.

Watching them, seeing how their lives are integrated into the rhythms of the Earth as it loops around the sun, makes me vividly aware of my own connection to the natural world and all that lives in it. You can feel the year unfolding, sense the orbit of the planet in the lifecycles of other species, exquisitely tuned to subtle changes in light and temperature.

24

TILTING
THE BALANCE

The quickening sense of anticipation felt by many swift-watchers in late spring is becoming increasingly tinged with anxiety. And never more so than in 2018 when an initial early flurry of swift arrivals was followed by a fortnight of extremely sparse sightings. Swallows and house martins were missing, too. A terrible feeling of foreboding took hold that this was an *annus horribilis*. What disastrous event had overtaken them? Violent storms? Pesticide spraying of locusts on their migration routes? Rumours circulated but no one could explain the awful void.

Most of them came back eventually. Data collected by BirdTrack, an online recording system, showed the reality of the decline: not a dramatically terrible year, but the downward trajectory on the graphs continued. Then summer arrived with sunshine every day, reviving people's spirits; at least the swifts would have a strong chance of raising their young successfully.

Yet the unease remains. Many of the world's swifts feed over forests and this is a fast-shrinking habitat. Despite pledges made in 2014 to halt the decline of forests globally, deforestation rates accelerated by 43 per cent in the following five years. That means 64 million acres of forest lost – most of it precious, primary rainforest. Forest-burning in the Amazon has intensified with the Brazilian president's active encouragement. Less widely known is that the rate of destruction in Central and West Africa has also rapidly increased. Huge losses have occurred in the common swift's main wintering areas, the Democratic Republic of the Congo, where more than two-thirds of Africa's intact rainforest is found. In a single year 1.4 million hectares of forest destruction was recorded. So-called sustainability regulation, which would limit the frequency of logging, is being flouted and the forestry companies, operating within secretive networks of shell companies, are inflicting ecological destruction on a massive scale, virtually unchallenged. No wilderness on Earth is now so wild and remote that it cannot be exploited and destroyed.

The international response to this catastrophic loss is utterly inadequate. Just 2 per cent of the funding for climate action goes towards forest and land protection, despite our heightened awareness of their value. As Frances Seymour at the World Resources Institute says, this is like 'trying to put out a house fire with a teaspoon'. Globally, forests absorb around a third of human-generated carbon

emissions; cutting them down exacerbates the problem.
The implications are disastrous both for people and the
vast array of wild creatures that thrive in and around them
– including swifts.

We use our world so recklessly. Ultimately, the consequences will come home, most directly through climate
change. Eroding soils and diminishing populations of polli-
nating insects threaten our food crops; global insect
abundance has shrivelled by 40 per cent in the last thirty-five
years. State of nature reports grow increasingly grim and
their frightening statistics ever harder to contemplate. No
longer, though, can we shrink back under the duvet of
wishful unknowing. Somehow, we have to grapple with
these stark realities.

At an individual level, how do we do this? 'What did
you do to save the planet?' my imaginary grandchildren
might ask. Not enough, I would say. I simply remembered
*Il faut cultiver notre jardin,** the closing words of Voltaire's
Candide, which I had studied at school, and set out to
do what I could in my garden. The novel was written in
1758, a fierce satire on the cosmic evils of the time, which
are little changed today. Taking up the decking and paving
slabs to let nature spiral forth was probably not quite
what the great philosopher had in mind when he wrote
those words, but it remains wise counselling and highly
relevant today. There is no better way to start dealing
with adversity than by working harmoniously within whatever space we occupy on Earth.

* we must cultivate our garden

Let sunlight touch the ground, give rainfall somewhere to soak away, grow something gorgeous which feeds the needs of wild creatures and nourishes your soul. As more and more of us live in towns and cities, we need to find ways of keeping nature in our lives: in gardens and window boxes, on rooftops, along our streets and around our workplaces. Such pleasure can be drawn from these interactions, renewing our tattered relationship with nature, refreshing our world and ourselves.

Moving beyond our own patch, we need to make our voice heard, lobby politicians, support the organisations whose staff risk their lives to expose the realities of environmental destruction, and make careful buying choices. In a global economy every purchase has ramifications; shopping has become an ethical minefield.

The support we give environmental charities is crucial. Many of them operate at a local level, wielding a positive influence within the immediate radius of our lives. Vigorous and sustained campaigning by organisations such as the Wildlife Trusts, RSPB and Friends of the Earth has influenced the shape of environmental legislation both in the UK and the EU. Ambitious landscape-scale conservation projects, involving partnerships of local organisations, agencies and individuals, would not happen without their leadership. It is argued by some that they do not do enough, that they have failed to stem the losses. Given the scale and multiplicity of environmental problems, this is scarcely surprising. Such organisations cannot save the world by themselves. Without them, however, things would be very much worse.

Meanwhile, it is a simple matter to provide swifts with a safe, dark hole where they can raise their young. And it is vital that we do so. It has been estimated that we need to integrate swift nesting bricks into 20,000 new houses each year for the next twenty years to restore an adequate number of nesting places. The availability of well-designed, cheap swift bricks makes this entirely possible. Swifts will ignore a high proportion of them but they will be used by other wildlife, too. Awareness of their requirements is growing and goodwill for swifts abounds, though it is not yet equally matched by active commitment.

My local swift group has constructed and installed a dozen nest boxes in two church towers in the town: one in the twelfth-century belfry of St Oswald's, with the enthusiastic blessing of the tower captain, the other in the louvred steeple of a Victorian church designed by the same architect as Llanfyllin's workhouse. The bells have gone from this tower now, but swift calls ring out each summer from morning to evening. We have seen them circling both towers but so far have found absolutely nothing in the way of a nest. I remind myself of other churches to which swifts have been attracted. The most successful is St Mary's Church, Ely, which now has a colony of more than forty pairs in boxes installed over the last twenty years or so. It can take anything from five minutes to fifteen years for swifts to find and occupy boxes. Sheltered from wind and rain inside the towers, wooden boxes will last for decades. Instant occupancy is gratifying but seldom occurs.

This summer brought some good stories about swifts. In Carmarthenshire, Wales, a man witnessed more than a dozen swifts trying and failing to enter their nest holes in a local chapel, where scaffolding had been erected for renovation work. The planks were blocking their access. Deeply disturbed by what he had seen, he determinedly set about trying to save them. He contacted the chapel's minister, who immediately pledged her commitment to look after the swifts, which she had not known were present. High-level scaffolding planks were taken down immediately and plans drawn up, with help from several knowledgeable swifting people, to ensure that the birds' nesting holes would be incorporated into the new fascia boards. A simple intervention from someone attentive to their existence thus saved a whole colony of birds. Keeping our eyes open and having local swift champions to talk to when help is needed can make a big difference.

Later in the season I heard from a friend in Nantwich, Cheshire, of evening gatherings over the town of hundreds of swifts. Once a common sight over towns and villages across the country – and still in southern Europe – such awesome numbers are rarely seen in the UK these days.

And here at home something thrilling happened, too. As in previous years, I watched the exhilarating evening flights of swifts hurtling through the narrow passageway by my house, racing their wild circuits. One evening I heard a shrill, short scream as the birds fled past. It was coming from inside the house! Shortly afterwards, a swift paused close to the apex of the roof, by the narrow crack between the boards and the wall, clinging to the brickwork.

Three times it did so, between circuits. Then the next
time, as the light faded, it scrambled through the gap to
join its mate. At last, I had swifts under my roof. In their
first summer they bred; I caught the soft calls of their
young when a parent bird brought food. Possibly they had
been displaced from elsewhere, fortunately finding the
hole I had reopened the year before, or perhaps they were
new breeders.

For five years I have immersed myself in the swifts' world.
The more I watch and learn about their behaviour, the
more intense is my wonder and excitement at their lives.
Their unique ability to live almost entirely in the air, to
find sustenance, rest and even nesting materials on the
wing, to fly across the world, navigating their way around
storms and depressions, never lighting on tree, cliff or
ground, then back to the hole in the very building that
their sharp-clawed feet shuffled around the previous year
– these are mind-blowing skills.

I think of gatherings of young adult birds spiralling into
the sky as the sun sets, drifting with the wind, half asleep,
half alert; of the breeders, out of their nests at sunrise,
snapping up flies and spiders for their young, foraging and
feeding for the next sixteen hours; and of the nestlings,
eager for food. I think of how their body temperature
will drop and their metabolism slow down if the weather
is cold, to save energy and survive; how their feathers
grow in the darkness of their holes, the gradual transfor-
mation from tiny, naked hatchling into sleek, aerial acrobat
in around six weeks, ready to fly to Africa the moment

they leave the nest hole. I remember my swiftling and how its luck ran out so early in life.

The brevity of their summer stays enhances their hold on our hearts. The season is short, their bold, wild chases over the roofs and high-pitched screams a fleeting experience: they are a metaphor for life itself. We need to act now to ensure these birds will scythe across our skies forever; to keep them in our streets, to keep them in abundance and *common*. All of us can do something within the compass of our lives to help tilt the balance back in their favour. If the will to do it is there, it can be done.

ACKNOWLEDGEMENTS

Huge thanks to all the people who have generously shared their expertise and given time and warm hospitality. I would particularly like to thank those who have offered continuous support, providing interesting and knowledgeable answers to a stream of questions over the years:

Edward Mayer, Dick Newell, Stephen Fitt, Mauro Ferri, Ulrich Tigges, Bernard Genton, Clare Darlaston, Lyndon Kearsley, Derek Bromhall, Andrew Lack, Helen Lack, Roy Overall, Jesús Solana, Elena Portillo, Esperanza Portillo, Enric Fusté, Marcel Jacquat, Giovanni Buono, Erika Tomassetto, Mark Smyth, Brian Cahalane, Peter Cush, John and Georgina Young, James O'Neill, George Candelin, Gillian Westray, Judith Wakelam, James Wolstonecroft, John Hawkins, John Tucker, Terry Townshend, Tom Wall, Roy Dennis, Uwe Stoneman, Daniele Muir, Sir John Lister-Kaye and Ben Jones.

Graham Williams, for expertly translating chapters of Weitnauer's *Mein Vogel* from German into English.

Andrew Tullo, Jake Allsop, Peter Lloyd, Jules Perry, John Hughes, Brian Hogbin and Tim Bierley for proofing and commenting on the text.

James Macdonald Lockhart of Antony Harwood Ltd, who responded with instant enthusiasm to an early draft of this book, and who, as my agent, persuaded me to develop it further, offering wonderful words of encouragement along the way.

Myles Archibald, publishing director of William Collins, who did me the unbelievable honour of offering to publish it, and Hazel Eriksson, editor, whose perceptive comments and suggestions gave me the chance to improve the text.

David Hardwick, Rachel Bierley, Mary Gibson and my team at Shropshire Wildlife Trust, for their constant encouragement and support for my singular project. And apologies to all my friends for being a little reclusive over the last five years.

Great thanks, too, to the inspirational swift groups in the UK and around the world, who show how individual, local action can make a real difference, and whose dynamism gave me the impetus to do something for swifts myself.

While I have made every effort to get my facts right, it is possible that errors may have crept in. I accept full responsibility should this have occurred.

REFERENCES

Derek Bromhall, *Devil Birds: The Life of the Swift*, Hutchinson & Co, 1980

Bernard Genton and Marcel Jacquat, *Martinet Noir: entre ciel et pierre*, Cahiers du MHNC No 15, publication conjointe du Cercle ornithologique des Montagnes neuchâteloises (COMON) et du Musée díhistoire naturelle de la Chaux-de-Fonds

David Lack, *Swifts in a Tower*, Methuen & Co Ltd, 1956

Emil Weitnauer, *Mein Vogel: Aus dem Leben des Mauerseglers*, 1980

Gilbert White, *The Natural History and Antiquities of Selborne*, Benjamin White, 1789

Chapter 1
Ted Hughes, 'Swifts', from *Season Songs*, Faber & Faber, 1976

Chapter 2

BTO Breeding Bird Survey recorded a 57 per cent decline between 1995 and 2017

Chapter 3

T. H. Huxley, 'On the Animals Which Are Most Nearly Intermediate Between Birds and Reptiles', *Annals and Magazine of Natural History*, 4th series, 2: 66–75, 1868

Gerald Lucy, 'The Geology of the Naze cliffs, Walton-on-the-Naze', geoessex.org.uk, revised May 2011

X. Xu and M. A. Norell, 'A New Troodontid Dinosaur From China with Avian-like Sleeping Posture', *Nature*, 431 (7010): 838–41, 2004

C. J. O. Harrison, 'A Revision of the Fossil Swifts (Vertebrata, Aves, suborder Apodi), With Descriptions of Three New Genera and Two New Species', *Meded. Werkgr. Tert. Kwart. Geol*, 21: 157–77, 1984

C. J. O. Harrison and C. Walker, 'A New Swift From the Lower Eocene of Britain', *Ibis*, 117: 162–4, 1975

D. T. Ksepka, J. A. Clarke, S. J. Nesbitt, F. B. Kulp and L. Grande, 'Fossil Evidence of Wing Shape in a Stem Relative of Swifts and Hummingbirds (Aves, Pan-Apodiformes)', *Proc. R. Soc. B.*, 280: 20130580, 2013

G. Mayr and D. S. Peters, 'On the Systematic Position of the Middle Eocene Swift *Aegialornis szarskii* Peters 1985 with Description of a New Swift-like Bird from Messel (Aves, Apodiformes), *N. Jb. Geol. Paläont. Mh.*, 312–20, 1999

G. Mayr, 'A New Eocene Swift-like Bird with a Peculiar Feathering', *Ibis*, 145, 382–91, 2003

Chapter 4

J. del Hoyo, N. J. Collar, D. A. Christie, A. Elliott and L. D. C. Fishpool, *HBW and BirdLife International Illustrated Checklist of the Birds of the World*, Volume I., Lynx Publications, 2014

Phil Chantler and Gerald Driessens, *Swifts: A Guide to the Swifts and Treeswifts of the World*, Pica Press, 2000

M. Päckert, A. Feigl, M. Wink and D. T. Tietze, 'Molecular Phylogeny and Historical Biogeography of Swifts (Apodidae: *Apus, Tachymarptis*)', 5th IBS Conference, 2011

Lisa Nupen, 'The Dazzling Diversity of Avian Feet', *African Bird Life*, Sept/October 2016

Rich Levad, *The Coolest Bird: The Natural History of the Black Swift and Those Who Have Pursued It*, American Birding Association, 2010

Lord Medway, 'The Swiftlets (*Collocalia*) of Niah Cave, Sarawak', Department of Zoology, University of Malaya in Kuala Lumpur, 1961

F. Liechti et al, 'First Evidence of a 200-day Non-stop Flight in a Bird', *Nat. Commun.*, 4 (2554), 2013

M. Ausden, R. Bradbury, A. Brown, M. Eaton, L. Lock and J. Pearce-Higgins, 'Climate Change and Britain's Birdlife: What Might We Expect?', *British Wildlife*, 26 (3), 2015

Chapter 5

A. M. Dokter, S. Åkesson, H. Beekhuis, W. Bouten, L. Buurma, H. van Gasteren and I. Holleman, 'Twilight Ascents by Common Swifts, *Apus apus*, at Dawn and Dusk: Acquisition of Orientation Cues?', *Animal Behaviour* 85 (3), 545–52, 2013

G. Guerin, 'La Vitesse de Vol des Oiseaux et l'Aviation', *Rev. franc. d'Ornith.*,15: 74–9

Chapter 6

'Of the House-Swallow, Swift, and Sand-Martin'. By the Rev. Gilbert White, in Three Letters to the Hon. Daines Barrington, F. R. S., *Phil. Trans. R. Soc.*, 65, 1775

Gilbert White, *The Natural History and Antiquities of Selborne*, Benjamin White, 1789: Letter X (Clergyman anecdote); Letter XX1; Letter XXXVI

Henry Reeve, *An Essay on the Torpidity of Animals*, Longman, 1809

David Lack, *Life of the Robin*, H. F. & G. Witherby Ltd, 1965

'Some Observations on the Migration of Birds'. By the late Edward Jenner, M. D. F. R. S.; with an Introductory Letter to Sir Humphry Davy, Bart. Pres. R. S. by the Rev. G. C. Jenner, *Phil. Trans. R. Soc.*, 114, 1824

Chapter 7

D. Lentink, U. K. Müller, E. J. Stamhuis, R. de Kat, W. van Gestel, L. Veldhuis, P. Henningsson, A. Hedenström, J. J. Videler and J. L. van Leeuwen, 'How Swifts Control Their Glide Performance with Morphing Wings', *Nature*, 446 (7139), 2007

J. Bäckman and T. Alerstam, 'Confronting the Winds: Orientation and Flight Behaviour of Roosting Swifts, *Apus apus*', *Proceedings: Biological Sciences*, 268 (1471), 2001

P. Henningsson, L. Christoffer Johansson and A. Hedenström, 'How Swift are Swifts *Apus apus?*', *J. Avian Biol.*, 41: 94–8, 2010

Hedenström et al, 'Annual 10-Month Aerial Life Phase in the Common Swift *Apus apus*', *Current Biology*, 26: 3066–70, 2016

Jan Holmgren, 'Roosting in Tree Foliage by Common Swifts *Apus apus*', *Ibis*, 146 (3): 404–16, 2004

Wall roosting observation by Julio Carralero: www.seomalaga.org

Chapter 8
The Museum's Architecture, Oxford University Museum of Natural History, oumnh.ox.ac.uk/museums-architecture

Christopher Perrins, 'Age of First Breeding and Adult Survival Rates in the Swift', *Bird Study*, 18 (2): 61–70, 1971

R. Overall, 'Guardian of the Swifts in the Tower of the Oxford University Museum of Natural History', *Fritillary*, 6, 2015

Chapter 9
Mark David Walker, 'An Investigation into the Host–Parasite Interrelationship Between Common Swifts and Hippoboscid Louse-Flies', Sheffield Hallam University, 2011

Chapter 11
T. L. F. Martins, J. K. Blakey and J. Wright, 'Low Incidence of Extra-pair Paternity in the Colonially Nesting Common Swift *Apus apus*', *Journal of Avian Biology*, 33 (4), 2002

Chapter 12
E. Fusté, E. Obon and L. Olid, 'Hand-Reared Common Swifts (*Apus apus*) in a Wildlife Rehabilitation Centre: Assessment of Growth Rates Using Different Diets', Centre de Recuperació de Fauna Salvatge de Torreferrussa, Barcelona, Spain

Chapter 13
S. Åkesson, R. Klaassen, J. Holmgren, J. W. Fox and A. Hedenström, 'Migration Routes and Strategies in a Highly Aerial Migrant, the Common Swift *Apus apus*, Revealed by Light-level Geolocators', PLOS ONE 7(7): e41195, 2012

Graham Appleton, 'Swifts Start to Share Their Secrets', *BTO News*, May–June 2012

'Out of Africa! The Beijing Swift's Incredible Journey Charted At Last': birdingbeijing.com

Chapter 14
Professor Dave Goulson, FRES, 'Insect Declines and Why They Matter', South West Wildlife Trusts, 2019

R. Al-Jaibachi, R. N. Cuthbert and A. Callaghan, 'Up and Away: Ontogenic Transference as a Pathway for Aerial Dispersal of Microplastics', *Biology Letters*, 14 (9), 2018

Chapter 15
John S. Wilson, 'Action to Halt the Decline of Swifts in Scotland', Report, 2012

Chapter 17
Sarah Roberts, *The Attitudes of Housing Occupants to Integral Birds and Bat Boxes*, Gloucester University, 2017

National Planning Policy Guidelines (NPPG 21/07/2019, paragraph 23)

Dr Carol Williams, Brian Murphy and Kelly Gunnell, *Designing for Biodiversity: A Technical Guide for New and Existing Buildings*, RIBA Publishing, 2012

Chapter 18
Konrad Ansorge, 'Sexual dimorphism of acoustic signals in the Common Swift *Apus apus*', *Apus Life* 2015, no 5457

Chapter 19
Mauro Ferri, 'Ancient Artificial Nests to Attract Swifts, Sparrows and Starlings to Exploit Them as Food', in F. Duhart and H. Macbeth, *Birds as Food: Anthropological and Cross-disciplinary Perspectives*, International Commission on the Anthropology of Food and Nutrition (2018)

Lazzaro Spallanzani, Booklet on the Common Swift, part of the six-volume *Viaggi alle due Sicilie e in alcune parti dell'Appennino* (1792–7); A personal communication from Mauro Ferri with translation support from Edward and Mandy Mayer

Chapter 23
E. Kaiser, 'Multiple Occurrence of Second Broods in Common Swift *Apus apus*', *Vogelwelt*, 125: 113–15, 2004

Chapter 24
Total Systems Failure: Exposing the Global Secrecy Destroying the Forests in the Democratic Republic of Congo, Global Witness, 2018: globalwitness.org

USEFUL WEBSITES
commonswift.org
actionforswifts.blogspot.com
swift-conservation.org

IMAGE CREDITS

The swift icons that appear on chapter openers are copyright Shutterstock.

Page   Credit

INDEX

Page references in *italics* indicate images.